NUMERICAL ANALYSIS, ALGORITHMS AND COMPUTATION

J. MURPHY, Ph.D.
Head of Department of Computational Physics
Sowerby Research Centre, British Aerospace plc, Bristol

D. RIDOUT, Ph.D.
Department of Computer Science and Mathematics
University of Aston, Birmingham

and

BRIDGID McSHANE, B.Sc.
Department of Mathematics
Cheadle Hulme School, Cheshire

ELLIS HORWOOD LIMITED
Publishers · Chichester

Halsted Press: a division of
JOHN WILEY & SONS
New York · Chichester · Brisbane · Toronto

First published in 1988 by
ELLIS HORWOOD LIMITED
Market Cross House, Cooper Street,
Chichester, West Sussex, PO19 1EB, England
The publisher's colophon is reproduced from James Gillison's drawing of the ancient Market Cross, Chichester.

Distributors:

Australia and New Zealand:
JACARANDA WILEY LIMITED
GPO Box 859, Brisbane, Queensland 4001, Australia

Canada:
JOHN WILEY & SONS CANADA LIMITED
22 Worcester Road, Rexdale, Ontario, Canada

Europe and Africa:
JOHN WILEY & SONS LIMITED
Baffins Lane, Chichester, West Sussex, England

North and South America and the rest of the world:
Halsted Press: a division of
JOHN WILEY & SONS
605 Third Avenue, New York, NY 10158, USA

South-East Asia
JOHN WILEY & SONS (SEA) PTE LIMITED
37 Jalan Pemimpin # 05–04
Block B, Union Industrial Building, Singapore 2057

Indian Subcontinent
WILEY EASTERN LIMITED
4835/24 Ansari Road
Daryaganj, New Delhi 110002, India

© 1988 J. Murphy, D. Ridout and B. McShane/Ellis Horwood Limited

British Library Cataloguing in Publication Data
Murphy, J.
Numerical analysis, algorithms and computation
1. Applied mathematics. Numerical analysis. Computation
I. Title   II. Ridout, D. (Dennis), *1934–*
III. McShane B., (Brigid), *1962–*   IV. Series
519.4

Library of Congress CIP available

ISBN 0–85312–751–4 (Ellis Horwood Limited, Library Edn.)
ISBN 0–7458–0549–3 (Ellis Horwood Limited, Student Edn.)
ISBN 0–470–21214–4 (Halsted Press)

Printed in Great Britain by Hartnolls, Bodmin

NUMERICAL ANALYSIS,
ALGORITHMS AND COMPUTATION

DATE DUE

MATHEMATICS AND ITS APPLICATIONS

Series Editor: G. M. BELL, Professor of Mathematics,
King's College London (KQC), University of London

NUMERICAL ANALYSIS, STATISTICS AND OPERATIONAL RESEARCH

Editor: B. W. CONOLLY, Professor of Mathematics (Operational Research),
Queen Mary College, University of London

Mathematics and its applications are now awe-inspiring in their scope, variety and depth. Not only is there rapid growth in pure mathematics and its applications to the traditional fields of the physical sciences, engineering and statistics, but new fields of application are emerging in biology, ecology and social organization. The user of mathematics must assimilate subtle new techniques and also learn to handle the great power of the computer efficiently and economically.

The need for clear, concise and authoritative texts is thus greater than ever and our series will endeavour to supply this need. It aims to be comprehensive and yet flexible. Works surveying recent research will introduce new areas and up-to-date mathematical methods. Undergraduate texts on established topics will stimulate student interest by including applications relevant at the present day. The series will also include selected volumes of lecture notes which will enable certain important topics to be presented earlier than would otherwise be possible.

In all these ways it is hoped to render a valuable service to those who learn, teach, develop and use mathematics.

Mathematics and its Applications

Series Editor: G. M. BELL, Professor of Mathematics, King's College London (KQC), University of London

Anderson, I.	Combinatorial Designs
Artmann, B.	The Concept of Number
Arczewski, K. & Pietrucha, J.	Mathematical Modelling in Discrete Mechanical Systems
Arczewski, K. and Pietrucha, J.	Mathematical Modelling in Continuous Mechanical Systems
Bainov, D.D. & Konstantinov, M.	The Averaging Method and its Applications
Baker, A.C. & Porteous, H.L.	Linear Algebra and Differential Equations
Balcerzyk, S. & Joszefiak, T.	Commutative Rings
Balcerzyk, S. & Joszefiak, T.	Noetherian and Krull Rings
Baldock, G.R. & Bridgeman, T.	Mathematical Theory of Wave Motion
Ball, M.A.	Mathematics in the Social and Life Sciences: Theories, Models and Methods
de Barra, G.	Measure Theory and Integration
Bartak, J., Herrmann, L., Lovicar, V. & Vejvoda, D.	Partial Differential Equations of Evolution
Bell, G.M. and Lavis, D.A.	Co-operative Phenomena in Lattice Models, Vols. I & II
Berkshire, F.H.	Mountain and Lee Waves
Berry, J.S., Burghes, D.N., Huntley, I.D., James, D.J.G. & Moscardini, A.O.	Mathematical Modelling Courses
Berry, J.S., Burghes, D.N., Huntley, I.D., James, D.J.G. & Moscardini, A.O.	Mathematical Methodology, Models and Micros
Berry, J.S., Burghes, D.N., Huntley, I.D., James, D.J.G. & Moscardini, A.O.	Teaching and Applying Mathematical Modelling
Blum, W.	Applications and Modelling in Learning and Teaching Mathematics
Brown, R.	Topology
Burghes, D.N. & Borrie, M.	Modelling with Differential Equations
Burghes, D.N. & Downs, A.M.	Modern Introduction to Classical Mechanics and Control
Burghes, D.N. & Graham, A.	Introduction to Control Theory, including Optimal Control
Burghes, D.N., Huntley, I. & McDonald, J.	Applying Mathematics
Burghes, D.N. & Wood, A.D.	Mathematical Models in the Social, Management and Life Sciences
Butkovskiy, A.G.	Green's Functions and Transfer Functions Handbook
Cartwright, M.	Fourier Methods: Applications in Mathematics, Engineering and Science
Cerny, I.	Complex Domain Analysis
Chorlton, F.	Textbook of Dynamics, 2nd Edition
Chorlton, F.	Vector and Tensor Methods
Cohen, D.E.	Computability and Logic
Cordier, J.-M. & Porter, T.	Shape Theory: Categorical Methods of Approximation
Crapper, G.D.	Introduction to Water Waves
Cross, M. & Moscardini, A.O.	Learning the Art of Mathematical Modelling
Cullen, M.R.	Linear Models in Biology
Dunning-Davies, J.	Mathematical Methods for Mathematicians, Physical Scientists and Engineers
Eason, G., Coles, C.W. & Gettinby, G.	Mathematics and Statistics for the Bio-sciences
El Jai, A. & Pritchard, A.J.	Sensors and Controls in the Analysis of Distributed Systems
Exton, H.	Multiple Hypergeometric Functions and Applications

Series continued at back of book

Table of Contents

1 Introduction   **1**

  1.1 Numerical/computational modelling   1
  1.2 Number representation and error   1
  1.3 Some prerequisites from Calculus   3

2 Iterative methods for non-linear equations   **7**

  2.1 Introduction   7
  2.2 Simple iteration by algebraic transformation   8
  2.3 Test for convergence   9
  2.4 Rate of convergence   14
  2.5 The bisection method   15
  2.6 The Newton–Raphson method for solving $f(x) = 0$ 17
  2.7 Equations with nearly equal roots   20
  2.8 Real roots of polynomial equations   22
      Exercises 2   26

3 Interpolation   **30**

  3.1 Finite differences and difference operators   32
  3.2 Interpolation formulae involving forward and backward diferences 40
  3.3 Divided differences and Newton's divided difference formula   45
  3.4 Central difference interpolation formulae   52
  3.5 Choices of interpolation polynomial   59
      Exercises 3   60

4 Numerical integration   **63**

  4.1 Approximation of the integrand   65
  4.2 Some well-known integration formulae   66
     4.2.1 The Trapezium rule   66

4.2.2 Simpson's rule   70
4.3 Truncation error and classification of integration formulae   75
4.4 Higher order integration formulae   76
4.5 Choice of formula   77
4.6 Romberg integration   78
4.7 Closed and open integration formulae   82
    Exercises 4   84

5 Numerical solution of ordinary differential equations   **87**

5.1 Multi-step methods   89
    5.1.1 Methods based on open integration   89
    5.1.2 Methods based on closed integration   92
    5.1.3 Starting methods   94
    5.1.4 More general multi-step methods   102
    5.1.5 Accuracy of multi-step formulae   103
    5.1.6 Systems of equations and higher order equations   110
5.2 Single-step methods   113
    5.2.1 First-order equations   113
    5.2.2 Systems of equations and higher order equations   118
    Exercises 5

6 Systems of linear equations   **120**

6.1 Direct methods   121
    6.1.1 Triangular systems—forward and backward substitution   121
    6.1.2 Gaussian elimination   122
    6.1.3 Triangular factorization   128
6.2 Iterative techniques   135
    6.2.1 Jacobi's method   135
    6.2.2 The Gauss–Siedel method   136
    6.2.3 Successive over-relaxation (S.O.R.)   137
    6.2.4 Convergence   139
    6.2.5 The iterative improvement procedure for removing round-off
          error   139
          Exercises 6   141

Appendix   **144**
    Homogeneous linear difference equations   144

Index   **147**

Preface

Numerical analysis and computation are important parts of many degree programmes in mathematics, engineering and science. This text is suitable for first- and second-year undergraduates following programmes within these subject areas. The material for this text has arisen from courses given to first- and second-year undergraduates in mathematics, engineering and science. To make the text more acceptable to students we have tried to minimize the number of advanced mathematical concepts. We have also attempted to present the more difficult concepts in a way that is understandable. The clarity has been enhanced by involving an author who has recently graduated and is closer to the difficulty of meeting this material for the first time. However, it is assumed that the reader will have a good knowledge of Calculus. This mean-value theorem and Taylor's theorem are of particular importance and are presented in Chapter 1.

An understanding of the introductory material in Chapter 1 is essential for almost the entire text. With the exception of Chapter 1, Chapters 2 and 6 are independent of the rest of the text. However, students who are studying this subject for the first time should read Chapters 3, 4 and 5 in that order because of the strong dependence of the later chapters on these early chapters. The material in Chapter 6 will be more easily understood by students who have attended an introductory course on linear algebra. Familiarity with the notation and some of the ideas of linear algebra is a distinct advantage. To allow time for a parallel course in linear algebra to take place we have put this material at the end of the text. The material of Chapter 6 can be brought forward for second-year students who have studied linear algebra in their first year.

Our treatment of interpolation, numerical integration and differential equations is strongly dependent on difference operators. This approach allows a smooth transition through Chapters 3, 4 and 5. A knowledge of finite difference calculus can be valuable to practising scientists and engineers involved in mathematical and numerical modelling—except those who prefer the black-box approach. There is a thin line between making the text acceptable to students and not compromising the mathematical rigour. This is particularly so in the numerical solution of differential equations, which deserves a separate text. We have derived some of the better known methods and given a relatively simple treatment of errors and stability. A more thorough treatment is beyond the intended scope of this text.

Since this is an introductory text, many of the problems and exercises can be solved with a hand calculator. However, numerical analysis cannot be separated from computation. The application of numerical analysis to a given problem usually leads to a numerical scheme or algorithm which is a sequence of arithmetic operations that leads to the solution of the problem. In many cases it is preferable or essential for these numerical algorithms to be implemented on a computer. Consequently computational efficiency influences trends in numerical analysis. Within the limits of presenting introductory material we have tried to emphasize this point throughout the text and have provided a number of algorithms with a Pascal structure. For clarity the algorithms are separated from the text throughout by being ruled off. Students are encouraged to enlarge these algorithms to a full program, either in Pascal or in another scientific language, and to test them on the examples.

ACKNOWLEDGEMENTS

The authors are extremely grateful to Mrs E. P. Bailey for her patience and commitment whilst typing the manuscript with its tedious mathematical notation.

1

Introduction

1.1 NUMERICAL/COMPUTATIONAL MODELLING

To investigate problems that arise in science, technology and commerce it is helpful to construct models which approximate the real situation. We are not referring to tangible models but to mathematical/computational models that give an approximate description of the behaviour of a real system. These models are usually formed by considering the important features or physical processes occurring in the real system and incorporating them into a system of equations, which may be algebraic but is often differential. The unknown quantities that we are seeking and that relate to the behaviour of the real system appear in these equations. Consequently the equations must be solved to determine these unknowns.

Real systems vary in their complexity. Although we have no control over this we do have control over the complexity of the models constructed to approximate the real systems. Simple systems can be approximated by simple models and in some cases the associated equations can be solved analytically. Complex systems can also be approximated by simple models but the models may not give an accurate description of the behaviour of the real system. A more accurate description of the real system may be obtained by incorporating more features of the real system into the model. Not surprisingly, the associated system of equations may be difficult to solve. Analytic solutions may not exist and the only way forward is to seek approximate solutions via numerical and computational methods. An introduction to these methods is given in this text.

1.2 NUMBER REPRESENTATION AND ERROR

The discussion that follows is equally applicable to positive and negative numbers, but for convenience in notation we discuss positive numbers only.

The decimal numbers that we work with can be expressed in the form

$$a = a_n 10^n + a_{n-1} 10^{n-1} + \cdots + a_m 10^m = \sum_{k=n(-1)m} a_k 10^k, \qquad (1.1)$$

where $n > m$ and a_k, $k = n(-1)m$, is an integer between 0 and 9. The notation we commonly use is

$$a = a_n a_{n-1} a_{n-2} \cdots a_0 \cdot a_{-1} a_{-2} \cdots a_m. \qquad (1.2)$$

Thus $n - m + 1$ digits are required to represent such a number. The number $n - m + 1$ can be extremely large and is infinite for many numbers, e.g. $1/3$, π, $\sqrt{2}$. Obviously, there is a limit on the magnitude of $n - m + 1$ when storing a number on a calculator or digital computer. This unavoidable error is called round-off error.

The restriction on the magnitude of numbers stored on computers and calculators would be quite severe if the form (1.2) was used. This restriction can be eased by expressing a real number as a number between zero and unity multiplied by a power of ten:

$$a = b \times 10^N (0 \leqslant b < 1). \qquad (1.3)$$

When a number is written in this form, b is called the mantissa and N is called the characteristic. This form can be obtained from (1.1) by taking 10^N, where $N = n + 1$, as a factor out of (1.1). This gives

$$a = (a_n 10^{-1} + a_{n-1} 10^{-2} + \cdots + a_m 10^{m-N}) 10^N.$$

Changing to a more convenient notation and noting that m can approach minus infinity we write

$$a = 0 \cdot d_1 d_2 d_3 \cdots d_k d_{k+1} \cdots \times 10^N, \qquad (1.4)$$

where N is a positive or negative integer, d_1 is an integar lying between one and nine and d_k, $k = 2, 3, 4, \ldots$, is an integer lying between zero and nine. Comparing (1.4) with (1.3) we see that the mantissa is

$$b = 0 \cdot d_1 d_2 d_3 \cdots d_k d_{k+1} \cdots. \qquad (1.5)$$

When the number, presented in the form (1.4), is stored on a computer, only a finite number of digits can be used to represent the mantissa. Thus the sequence of digits representing the mantissa must be terminated, and this leads to the floating-point form of a number which we denote by $fl(a)$. There are two ways of doing this. One method is simply to chop all digits after the kth digit d_k, giving

$$fl(a) = 0 \cdot d_1 d_2 d_3 \cdots d_k \times 10^N, \qquad (1.6a)$$

where k will depend on the computer being used. This method is described as chopping. The other method is to add 5 to the digit d_{k+1} and then chop to obtain

$$fl(a) = 0 \cdot \delta_1 \delta_2 \delta_3 \cdots \delta_k \times 10^N. \qquad (1.6b)$$

This method is called rounding. If $d_{k+1} < 5$, rounding is exactly equivalent to chopping and we say that a has been rounded down. If $d_{k+1} \geqslant 5$, the rounding process adds one to d_k to produce δ_k and we say that a has been rounded up. When $d_k = 9$, rounding up produces $\delta_k = 0$ and adds one to d_{k-1} to produce δ_{k-1}. Similarly, if $d_{k-1} = 9$ as well as

$d_k = 9$, we would also have $\delta_{k-1} = 0$ and one would be added to d_{k-2} to produce δ_{k-2} etc. This is the rounding process that most of us are familiar with.

Before carrying out any calculations we always need to specify the accuracy required for the results. This is normally done in terms of significant digits. The rounded number

$$a* = 0 \cdot c_1 c_2 c_3 \cdots c_k \times 10^N$$

gives an accurate representation of a to k significant digits if

$$a* = \text{fl}(a), \quad \text{i.e. } c_i = \delta_i, \quad i = 1(1)n. \tag{1.7}$$

To relate the number of significant digits to error we first need to define absolute error and relative error. If $a*$ is an approximation to a, the absolute error is $|a - a*|$ and the relative error is $|a - a*|/|a|$. Then, the number $a*$ is said to approximate a to k significant digits if k is the largest non-negative integer for which

$$\frac{|a - a*|}{|a|} < 5 \times 10^{-k}. \tag{1.8a}$$

For many applications, sufficient significant figures are kept when a real number is stored on a computer. However, rounding errors can cause difficulty with specific problems and special methods are usually needed (see section 6.2.5, for example). Many calculators decrease the effect of round-off error in the displayed result by using $D + 2$ significant figures for the internal arithmetic operations, where D is the number of digits displayed.

The inequality (1.8a) has a wide application. Its use is not limited to checking for round-off error. It simply relates the idea of significant figures to relative error, and the error in $a*$ could come from other sources such as truncation error (e.g. section 5.1.5). Many numerical procedures generate a sequence of approximations with the aim that the sequence converges to the true solution (e.g. Chapter 2). The inequality (1.8a) can be used to check for convergence of the sequence, i.e. if a_n and a_{n+1} are two elements of the sequence we require

$$\frac{|a_{n+1} - a_n|}{|a_{n+1}|} < 5 \times 10^{-k} \tag{1.8b}$$

for k-significant-figure accuracy.

1.3 SOME PREREQUISITES FROM CALCULUS

Although it is expected that the reader will have a good understanding of Calculus, we review two important results that are used throughout the text.

The mean-value theorem

Before describing the mean-value theorem we mention Rollé's theorem which can be regarded as a special case of the mean-value theorem. Both Rollé's theorem and the mean-value theorem are easily understood when considered geometrically.

Consider a function $y = f(x)$ and suppose that

(i) the function is continuous for $a \leqslant x \leqslant b$,

(ii) the function has a derivative at every value of x in the range $a < x < b$,
(iii) the values of the function at $x = a$ and $x = b$ are equal.

Then, Rollé's theorem states the geometrically obvious result that there exists at least one point between $x = a$ and $x = b$ where the derivative of the function is zero, i.e. there exists at least one point $x = c$ such that $a < c < b$ and $f'(c) = 0$. Geometrically this means that there must be a point $x = c$ between $x = a$ and $x = b$ where the tangent to the curve is parallel to the x-axis.

The mean-value theorem is an extension of Rollé's theorem which relaxes condition (iii) given above. Thus consider a function $y = f(x)$ and suppose that

(i) the function is continuous for $a \leqslant x \leqslant b$,
(ii) the function is differentiable for $a < x < b$.

Then, the mean-value theorem states that there exists at least one number c such that $a < c < b$ and

$$f'(c) = \frac{f(b) - f(a)}{b - a}.$$
(1.9)

The left-hand side of this equation gives the gradient of the tangent to the curve at $x = c$. The right-hand side gives the gradient of the chord joining the points $(a, f(a))$ and $(b, f(b))$. Thus, geometrically, the mean-value theorem states that, when conditions (i) and (ii) are satisfied, there must be at least one point $x = c$ between $x = a$ and $x = b$ where the tangent to the curve is parallel to the chord. This is shown in Fig. 1.1.

Although the mean-value theorem is intuitive geometrically we show that it can be derived by applying Rollé's theorem to the function

$$\phi_1(x) = F_1(x) - \frac{(b - x)}{(b - a)} F_1(a),$$
(1.10)

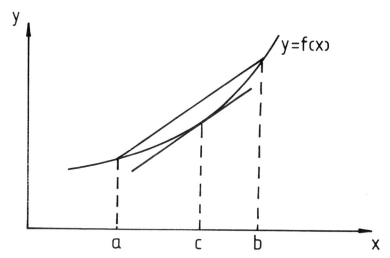

Fig. 1.1. Geometrical interpretation of the mean-value theorem. When conditions (i) and (ii) of the mean-value theorem are satisfied there is a point c, between a and b, where the tangent is parallel to the chord joining $(a, f(a))$ to $(b, f(b))$.

where

$$F_1(x) = f(b) - f(x).$$

The function $\phi_1(x)$ satisfies the conditions $\phi_1(a) = 0$ and $\phi_1(b) = 0$. Conditions (i), (ii) of the mean-value theorem require that $f(x)$ is continuous for $a \leqslant x \leqslant b$ and differentiable for $a < x < b$. Consequently $\phi_1(x)$ also satisfies these conditions, showing that the conditions for Rollé's theorem are satisfied by the function $\phi_1(x)$. Thus, applying Rollé's theorem to the function $\phi_1(x)$, there exists at least one value $x = c$, $a < c < b$, such that $\phi'_1(c) = 0$. From the definition of $\phi_1(x)$ we have

$$\phi'_1(x) = -f'(x) + \frac{f(b) - f(a)}{b - a}$$

and substituting $x = c$ gives equation (1.9).

Taylor series

This is a particularly important result in almost all work involving approximations. Because it can be used to obtain power series expansions of functions the error involved in approximations can be examined.

From equation (1.9) the mean-value theorem states that there exists at least one point $x = c$, $a < c < b$, such that

$$f(b) = f(a) + (b - a)f'(c), \tag{1.11}$$

provided that $f(x)$ is continuous for $a \leqslant x \leqslant b$ and has a derivative throughout the interval $a < x < b$. Written in this form, the mean-value theorem can be regarded as a simple Taylor expansion. Above, we derived the mean-value theorem by applying Rollé's theorem to a suitably chosen function, namely the function $\phi_1(x)$ defined in equation (1.10). We adopt a similar approach to derive more general Taylor expansions from Rollé's theorem. First, consider the function

$$\phi_2(x) = F_2(x) - \frac{(b - x)^2}{(b - a)^2} F_2(a)$$

where

$$F_2(x) = f(b) - f(x) - (b - x)f'(x).$$

Clearly $\phi_2(a) = \phi_2(b) = 0$ and, provided that $f'(x)$ is continuous for $a \leqslant x \leqslant b$ and that $f''(x)$ exists for $a < x < b$, the function $\phi_2(x)$ satisfies the conditions in Rollé's theorem. Thus, there exists a point $x = c$, $a < c < b$, such that $\phi'_2(c) = 0$. Further, differentiation gives

$$\phi'_2(x) = F'_2(x) + 2\frac{(b - x)}{(b - a)^2} F_2(a)$$

and

$$F'_2(x) = -f'(x) + f'(x) - (b - x)f''(x),$$
$$= -(b - x)f''(x)$$

leading to

$$\phi'_2(x) = 2\frac{(b - x)}{(b - a)^2} \{f(b) - f(a) - (b - a)f'(a) - \tfrac{1}{2}(b - a)^2 f''(x)\}.$$

Substituting $x = c$ leads to the result

$$f(b) = f(a) + (b - a)f'(a) + \tfrac{1}{2}(b - a)^2 f''(c) \tag{1.12}$$

where $a < c < b$.

Equations (1.11) and (1.12) are special cases of Taylor's theorem. To obtain the general result, consider the function

$$\phi_n(x) = F_n(x) - \frac{(b - x)^n}{(b - a)^n} F_n(a),$$

where

$$F_n(x) = f(b) - f(x) - (b - x)f'(x) - \cdots - \frac{(b - x)^n}{(n - 1)!} f^{(n-1)}(x).$$

Then $\phi_n(a) = \phi_n(b) = 0$ and, provided that the first $n - 1$ derivatives of $f(x)$ are continuous for $a \leqslant x \leqslant b$ and $f^{(n)}(x)$ exists for $a < x < b$, the function $\phi_n(x)$ satisfies the conditions of Rollé's theorem. Hence, there exists a point $x = c$, $a < c < b$, such that $\phi_n'(c) = 0$. Differentiating the above expression for $\phi_n(x)$ and substituting $x = c$ leads to, after some cumbersome algebra, the result known as Taylor's theorem,

$$f(b) = f(a) + (b - a)f'(a) + \frac{(b - a)^2}{2!} f''(a)$$

$$+ \cdots + \frac{(b - a)^{n-1}}{(n - 1)!} f^{(n-1)}(a) + \frac{(b - a)^n}{n!} f^n(c), \tag{1.13}$$

where $a < c < b$. This result is often expressed in the form

$$f(x) = f(x_0) + hf'(x_0) + \frac{h^2}{2!} f''(x_0) + \cdots + \frac{h^{n-1}}{(n - 1)!} f^{(n-1)}(x_0) + R_n(x), \tag{1.14}$$

where

$$R_n(x) = \frac{h^n}{n!} f^n(c) \tag{1.15}$$

and $x_0 < c < x$. This form is obtained directly from (1.13) by setting $x = b$, $x_0 = a$ and $h = b - a = x - x_0$. A slight variation on (1.14), that is commonly used, is

$$f(x_0 + h) = f(x_0) + hf'(x_0) + \frac{h^2}{2!} f''(x_0) + \cdots. \tag{1.16}$$

2

Iterative Methods for Non-Linear Equations

2.1 INTRODUCTION

Standard analytical techniques are of no use for solving equations such as

(i) $8.146x^5 - 3.219x^4 + 7.683x^3 + 4.17x + 5.139 = 0$
(ii) $x = 3 \sin x$.

We therefore need to develop methods for finding numerical approximations to the roots of such equations. The method employed is called *iteration*.

We concentrate on single non-linear equations in one unknown such as (i) or (ii). These equations can be expressed in the form

$$f(x) = 0. \tag{2.1}$$

Denoting an exact root of this equation by α, so that $f(\alpha) = 0$, iteration involves making an *initial guess*, x_0 say, to the root α and *generating a sequence*

$$x_1, x_2, x_3, \ldots, x_n, \ldots$$

from the initial guess, with the intention that the sequence converges to the root α. Below we discuss how to obtain the initial guess and compute the sequence. Convergence is achieved if, after sufficient elements of the sequence have been computed, the difference between x_n and α decreases as n increases. Convergence is detected when the difference between successive elements decreases. The computation is terminated when two successive elements x_n, x_{n+1} satisfy

$$|x_{n+1} - x_n| < \varepsilon,$$

where ε is a prescribed tolerance. The number n will depend on ε since, for a given computational scheme, increased accuracy is achieved by additional computation.

An initial guess to the root of (2.1) can be obtained by sketching the function $y = f(x)$ and estimating where it crosses the x-axis. An accurate graph is not needed. Sometimes

7

it is simpler to rearrange the equation $f(x) = 0$ into the form $f_1(x) = f_2(x)$ and to sketch each of $y_1 = f_1(x)$ and $y_2 = f_2(x)$. The intersection of the two curves gives the initial guess. For example, an initial guess to the root of

$$x^2 = e^x$$

is more easily obtained by sketching each of $y_1 = x^2$ and $y_2 = e^x$ rather than sketching $y = x^2 - e^x$.

From an initial guess, the sequence $x_1, x_2, \ldots, x_n, \ldots$ can be computed from an equation of the form

$$x_{n+1} = g(x_n). \tag{2.2}$$

Such an equation relating successive elements of a sequence is sometimes called a *recurrence relation*. We will discuss different ways of obtaining equations of the form (2.2).

2.2 SIMPLE ITERATION BY ALGEBRAIC TRANSFORMATION

In this section we introduce some of the ideas associated with iteration and show some of the difficulties that can arise. We attempt to overcome these difficulties in the sections that follow. The simple iterative technique described in this section are rarely used because of their poor convergence properties.

There are several algebraic rearrangements of equation (2.1) that take the form

$$x = g(x) \tag{2.3}$$

and since equations (2.1) and (2.3) have the same roots we may solve equation (2.3) to find the roots of (2.1). Later we will make use of the equation

$$f(\alpha) = 0 \quad \text{and} \quad \alpha = g(\alpha), \tag{2.4}$$

where $x = \alpha$ is one of the roots. Equations (2.3) and (2.2) have a similar form. Indeed, one of the ways of obtaining a recurrence relation of the form (2.2) is to rearrange $f(x) = 0$ into a form (there are usually several alternative forms) $x = g(x)$ and

(i) attach the suffix $n + 1$ to the single x on the left-hand side,
(ii) attach the suffix n to all x's on the right-hand side, i.e. in the function $g(x)$.

If the sequence converges, which we discuss later, we have found a root of $x = g(x)$.

We demonstrate some of the ideas associated with iteration by solving the simple quadratic equation

$$x^2 - 3x + 1 = 0. \tag{2.5}$$

Using the quadratic formula the roots of this equation are found to be 2.6180 and 0.3820. We will try to obtain these roots by iteration. One rearrangement of equation (2.5) that takes the form $x = g(x)$ is

$$x = \tfrac{1}{3}(x^2 + 1)$$

and comparison with (2.3) shows that $g(x) = (x^2 + 1)/3$ for this arrangement of the equation. The corresponding computational scheme is

$$x_{n+1} = \tfrac{1}{3}(x_n^2 + 1). \tag{2.6}$$

At this stage we try some different initial guesses and see what happens. Taking $x_0 = 1$ leads to

$$x_1 = \tfrac{1}{3}(1^2 + 1) = 0.667, \quad x_2 = \tfrac{1}{3}(0.677^2 + 1) = 0.481,$$
$$x_3 = 0.411, \quad x_4 = 0.390, \text{ etc.}$$

and this is converging to the smaller root. Starting with $x_0 = 2$, intuition may lead us to expect convergence to the larger root. However, the sequence

$$x_1 = \tfrac{1}{3}(2^2 + 1) = 1.667, \quad x_2 = 1.260, \quad x_3 = 0.863, \quad x_4 = 0.572, \ldots$$

arises, which is again converging to the smaller root. Clearly, intuition can be misleading. One might also expect the initial guess $x_0 = 3$ to lead to the larger root. In this case the reader may wish to check that the sequence obtained from (2.6) diverges.

Since it appears that the larger root cannot be obtained from equation (2.6) it is necessary to take an alternative rearrangement of equation (2.5). Taking $x = 3 - 1/x$, so that $g(x) = 3 - 1/x$, gives the scheme

$$x_{n+1} = 3 - \frac{1}{x_n}$$

and an initial guess of $x_0 = 1$ leads to the sequence $2.000, 2.500, 2.600, 2.615, \ldots$ which is converging to the larger root.

The trial and error method used in the above example is inefficient. Before carrying out any calculations it would be preferable to be able to test whether a given iterative scheme with initial guess will converge. We pursue this point in the following section.

2.3 TEST FOR CONVERGENCE

We look for conditions that will guarantee convergence of the scheme $x_{n+1} = g(x_n)$ from an initial guess x_0 to an exact root α of $f(x) = 0$. Since it is g that generates the sequence $x_1, x_2, x_3, \ldots$ we call g the *sequence generating function*. If this function varies rapidly with x there will be large differences between the elements of the sequence. Therefore, for convergence, it appears desirable to identify an interval that contains the root and for which $g'(x)$ has small magnitude. If this is not possible the sequence generating function is unsuitable and an alternative must be found. So we locate an interval $[a, b]$ containing the root α. An interval $[a, b]$ contains the root if $f(x)$ is continuous on $[a, b]$ and if $f(a)$ and $f(b)$ have opposite signs. To investigate convergence we concern ourselves with the behaviour of the quantity $|x_n - \alpha|$ as n increases and this will depend on the behaviour of the sequence generating function. Therefore we use the equations $x_n = g(x_{n-1})$ and $\alpha = g(\alpha)$ to obtain

$$|x_n - \alpha| = |g(x_{n-1}) - g(\alpha)|. \tag{2.7}$$

Above, we argued that convergence would depend on $g'(x)$ and, by using the mean-value theorem (Chapter 1), we can introduce $g'(x)$ into equation (2.7). To apply this theorem we require $g(x)$ to have a continuous derivative on the interval I_{n-1} defined by the end points x_{n-1} and α. Then, according to the mean value theorem, there exists a value of x in I_{n-1}, which we denote by c_{n-1}, such that

$$g'(c_{n-1}) = \frac{g(x_{n-1}) - g(\alpha)}{x_{n-1} - \alpha}.$$

Substituting into (2.7) yields

$$|x_n - \alpha| = |g'(c_{n-1})||x_{n-1} - \alpha|, \tag{2.8}$$

which relates the closeness of two successive approximations to the root α. If x_n is to be closer to the root than x_{n-1} we require that

$$|g'(c_{n-1})| < 1. \tag{2.9}$$

In a similar manner we can show that

$$|x_{n-1} - \alpha| = |g'(c_{n-2})||x_{n-2} - \alpha|.$$

Substituting into equation (2.8) yields

$$|x_n - \alpha| = |g'(c_{n-1})||g'(c_{n-2})||x_{n-2} - \alpha|.$$

Proceeding in this way, using (2.8) with n replaced by $n-2, n-3, \ldots, 1$ we obtain

$$|x_n - \alpha| = |g'(c_{n-1})||g'(c_{n-2})| \cdots |g'(c_0)||x_0 - \alpha|. \tag{2.10}$$

It is evident that the scheme will converge if

$$|g'(c_j)| < 1 \tag{2.11}$$

for $j = 0, 1, 2, \ldots, n-1$. Since we know little about the location of each c_j, except that it lies in I_j (recall I_j has end points α and x_j), the condition (2.11) will only be of use if

$$|g'(x)| < 1$$

over an interval sufficiently large to contain each interval I_j. Consequently, not only must we identify an interval $[a, b]$ containing α, but also the sequence generating function must satisfy the conditions

(i) $|g'(x)| < 1$ for all x in $[a, b]$
(ii) $a \leqslant g(x) \leqslant b$ for all x in $[a, b]$,   (2.12)

as well as having a continuous derivative on $[a, b]$. If the second condition is not satisfied, recalling that $x_j = g(x_{j-1})$, some of the x_j's may lie outside $[a, b]$. This allows the possibility of c_j, bounded by α and x_j, also lying outside $[a, b]$ and the first condition may be violated when $x = c_j$.

We have identified conditions that are sufficient for the scheme $x_{n+1} = g(x_n)$ to converge to a root α of $f(x) = 0$. To complete the discussion we will assume that conditions (i) and (ii) hold and confirm that convergence is guaranteed. Since $|g'(x)| < 1$ on $[a, b]$ we can find an upper bound K for $|g'(x)|$ on $[a, b]$ which is also less than one, e.g. $K = \max |g'(x)|$ or $K = (1 + \max |g'(x)|)/2$ where $a \leqslant x \leqslant b$. Then since

$$|g'(x)| \leqslant K < 1,$$

equation (2.10) leads to

$$|x_n - \alpha| \leqslant K^n |x_0 - \alpha|.$$

Furthermore, since $K < 1$, $K^n \to 0$ as $n \to \infty$ so that $x_n \to \alpha$ as $n \to \infty$, i.e. the scheme converges.

In summary, given an iterative scheme $x_{n+1} = g(x_n)$, an interval $[a, b]$ must be found containing the root α of $f(x) = 0$. The scheme will then converge to the root α provided that

(i) *The initial guess x_0 is selected from $[a, b]$,*
(ii) *g has a continuous derivative on $[a, b]$,*
(iii) *$|g'(x)| < 1$ for all x in $[a, b]$,*
(iv) *$a \leqslant g(x) \leqslant b$ for all x in $[a, b]$.*

When using the convergence test to decide on an iterative scheme an interval $[a, b]$ must be located which contains the root. The first step is to obtain an initial guess x_0 and this is usually achieved from a rough sketch. Allowing for the 'roughness' of the sketch, an interval $[a, b]$ can then be identified which is sufficiently large to contain the root. It remains to find a scheme satisfying conditions (ii), (iii), (iv). These conditions must be satisfied over the whole interval $[a, b]$. In particular it is not sufficient to show that $|g'(x)| < 1$ at a few isolated points such as the end points $x = a$, $x = b$ and possibly the mid-point $x = (a + b)/2$. The values produced by the sequence generating function may lie anywhere in $[a, b]$ and consequently it is necessary that $|g'(x)| < 1$ at every point of $[a, b]$.

The next example illustrates the above routine, and this is followed by its algorithmic description.

Example 2.1 Use a simple iterative method to solve

$$x^3 + 2x + 1 = 0.$$

By sketching the functions $y_1 = x^3$ and $y_2 = -2x - 1$ (Fig. 2.1) we see that the only real root lies close to -0.4 and in the interval $[-0.75, -0.25]$. One rearrangement of the

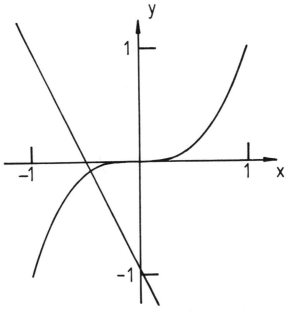

Fig. 2.1. To obtain initial approximations to the roots of $x^3 + 2x + 1 = 0$ the functions $y_1 = x^3$ and $y_2 = -2x - 1$ have been sketched. There is one real root near $x = -0.4$.

equation is

$$x = -(1 + 2x)/x^2$$

and for this form

$$g(x) = -(1 + 2x)/x^2, \quad g'(x) = 2(1 + 1/x)/x^2.$$

Since $g'(-0.4) > 1$, convergence is not guaranteed and we try the alternative rearrangement

$$x = -(1 + x^3)/2.$$

Here

$$g(x) = -(1 + x^3)/2, \quad g'(x) = -3x^2/2$$

and to have $|g'(x)| < 1$ we require that $3x^2/2 < 1$. This is the case if

$$x^2 - 2/3 < 0$$

$$(x - \sqrt{(2/3)})(x + \sqrt{(2/3)}) < 0.$$

This inequality is satisfied if the two factors on the left-hand side have opposite signs. This implies $x < \sqrt{(2/3)}$ and $x > -\sqrt{(2/3)}$. Hence $|g'(x)| \leqslant 1$ when x lies in the interval $[-\sqrt{(2/3)}, +\sqrt{(2/3)}]$. We can restrict attention to the interval $[-\sqrt{(2/3)}, 0]$ since it contains the root. Noting that the function $g(x)$ is monotonic (a function is monotonic on $[a, b]$ if it is always increasing or decreasing on $[a, b]$, i.e. its gradient does not change sign on $[a, b]$), its maximum and minimum values occur at the ends of the interval $[-\sqrt{(2/3)}, 0]$. These extreme values are -0.3519 and -0.5 and it is evident that $-\sqrt{(2/3)} < g(x) < 0$. Thus convergence of the scheme

$$x_{n+1} = -(1 + x_n^3)/2$$

is guaranteed provided an initial guess is taken from the interval $[-\sqrt{(2/3)}, 0]$. Taking $x_0 = -0.4$ we find that $x_6 = -0.45336$ and $x_7 = -0.45341$. Thus, to four decimal places the root is -0.4534.

Example 2.2 Show that the equation

$$(2x + 1)^2 = 4\cos \pi x$$

has a root in the interval $[1/4, 1/3]$ and determine whether either of the iterative schemes

$$x_{n+1} = \{(\cos \pi x_n)^{1/2} - 1/2\}, \quad x_{n+1} = \frac{1}{\pi}\cos^{-1}\{(x_n + \tfrac{1}{2})^2\}$$

will converge to the root for any starting value in this interval.
    If we set

$$f(x) = (2x + 1)^2 - 4\cos \pi x$$

we see that $f(1/4) < 0$ and $f(1/3) > 0$ and since $f(x)$ is continuous on $[1/3, 1/4]$ there must be a root of $f(x) = 0$ in this interval.
    For the first scheme

$$g(x) = (\cos \pi x)^{1/2} - 1/2, \quad g'(x) = -\frac{\pi \sin \pi x}{2(\cos \pi x)^{1/2}}$$

and at the mid-point of $[1/3, 1/4]$ we have $g'(0.2917) = 1.6$. Since $g'(x)$ is continuous there is an interval within $[1/3, 1/4]$ over which $|g'(x)| > 1$. Therefore convergence from any starting point within $[1/3, 1/4]$ is not guaranteed.

For the second scheme we have

$$g(x) = \frac{1}{\pi} \cos^{-1}\{(x + \tfrac{1}{2})^2\}, \quad g'(x) = \frac{2}{\pi} \frac{(x + 1/2)}{\sqrt{\{1 - (x + 1/2)^4\}}}$$

and clearly $g'(x) > 0$ on the interval $[1/4, 1/3]$. Consequently the maximum and minimum values of $g(x)$ will occur at the ends of this interval. These are $g(1/3) = 0.256$ and $g(1/4) = 0.310$ and therefore $g(x)$ certainly satisfies the condition $1/4 \leqslant g(x) \leqslant 1/3$ for any x in $[1/4, 1/3]$. To examine the behaviour of $g'(x)$ on the interval let

$$h(x) = \frac{2}{\pi} \frac{(x + 1/2)}{\sqrt{\{1 - (x + 1/2)^4\}}};$$

then

$$h'(x) = \frac{2}{\pi} \frac{\{1 + (x + 1/2)^4\}}{\{1 = (x + 1/2^4\}}$$

which is positive on $[1/4, 1/3]$. Therefore $h(x)$ is monotonic on $[1/4, 1/3]$ and its maximum and minimum values occur at the end points. These are $h(1/4) = 0.577$ and $h(1/3) = 0.737$. Consequently $|g'(x)| < 1$ over the entire interval $[1/4, 1/3]$ and convergence is guaranteed.

This last result can also be shown algebraically. The result $|g'(x)| < 1$ will hold when x satisfies the inequality

$$\frac{2}{\pi}(x + 1/2) < \sqrt{(1 - (x + 1/2)^4)}$$

and for the range of values of x under consideration

$$\frac{4}{\pi^2} y^2 < 1 - y^4,$$

where we have set $y = x + 1/2$ to simplify the algebra. Hence

$$y^4 + \frac{4}{\pi^2} y^2 - 1 < 0$$

$$(y^2 - 0.818)(y^2 + 1.222) < 0$$

implying that

$$y^2 - 0.818 < 0$$

$$(y - 0.904)(y + 0.904) < 0.$$

This is satisfied if

$$y - 0.904 > 0, \quad y + 0.904 < 0$$

or

$$y - 0.904 < 0, \quad y + 0.904 > 0.$$

The first is not possible and the second condition requires that

$$-1.404 < x < 0.504.$$

This certainly includes the interval $[1/4, 1/3]$ and therefore $|g'(x)| < 1$ on this interval.

Starting with the mid-point of $[1/4, 1/3]$ as initial guess the second scheme requires 15 iterations to converge to the root 0.2872 with four-decimal-place accuracy.

Convergence of the iterative scheme in the above example was rather slow and with some schemes it can be even slower. Clearly it would be useful to have some idea of the rate of convergence of an iterative scheme before using it. We investigate this further in the next section.

Algorithm 2.1. The iterative scheme based on the form $x = g(x)$

Both the initial approximation x and $g(x)$ must lie in an interval throughout which $|g'(x)| < 1$; Tol is the tolerance.

```
read(x, Tol);
k:= 0;
REPEAT
    k:= k + 1;
    x01d:= x;
    x:= g(x)
    UNTIL abs(x − x01d) < Tol;
write(x, k).
```

2.4 RATE OF CONVERGENCE

Now that we know how to determine whether a given iterative scheme is convergent it seems natural to ask the question 'Which of the convergent schemes will converge most rapidly?' Thus we need some means of grouping the convergent schemes according to their rate of convergence. To do this we will compare the error at successive stages of an iterative process.

Consider the iterative scheme $x_{n+1} = g(x_n)$ for solving $f(x) = 0$ and let α be an exact solution so that $f(\alpha) = 0$ and $\alpha = g(\alpha)$. The error, ε_n, at the nth stage is defined by the equation

$$x_n = \alpha + \varepsilon_n. \tag{2.13}$$

To relate the error at successive stages we simply substitute (2.13) into the recurrence relation, giving

$$\alpha + \varepsilon_{n+1} = g(\alpha + \varepsilon_n)$$

and provided $g(x)$ is sufficiently differentiable we can make use of a Taylor expansion

$$\alpha + \varepsilon_{n+1} = g(\alpha) + \varepsilon_n g'(\alpha) + \frac{\varepsilon_n^2}{2!} g''(\alpha) + \frac{\varepsilon_n^2}{3!} g^{(3)}(\alpha) + \cdots.$$

Then since $\alpha = g(\alpha)$ we have

$$\varepsilon_{n+1} = \varepsilon_n g'(\alpha) + \frac{\varepsilon_n^2}{2!} g''(\alpha) + \frac{\varepsilon_n^3}{3!} g^{(3)}(\alpha) + \cdots. \tag{2.14}$$

Since we are only considering convergent schemes we know that $\varepsilon_n \to 0$ as $n \to \infty$ and for sufficiently large n the right-hand side of (2.14) can be approximated by the *first non-zero term*. Hence if $g'(\alpha) \neq 0$ we have

$$\varepsilon_{n+1} \simeq \varepsilon_n g'(\alpha).$$

Here the rate of convergence is dominated by the *first* derivative of g, and ε_{n+1} is proportional to the *second* power of ε_n. Consequently schemes for which the *first non-vanishing derivative of g is the second* are called *second-order schemes*.

In general an *iterative scheme* $x_{n+1} = g(x_n)$ *is said to have order m when the first non-vanishing derivative of g at $x = \alpha$ is the mth.* Equation (2.14) then approximates to

$$\varepsilon_{n+1} \simeq \frac{\varepsilon_n^m}{m!} g^m(\alpha). \tag{2.15}$$

Obviously higher order schemes converge more rapidly. However the formulae associated with such schemes are often more complicated, and the more rapid convergence may be partly offset by additional computation at each stage of the iteration. Only first- and second-order processes are normally used.

Example 2.3 Determine the order of the convergent scheme in Example 2.2.
For the scheme in Example 2.2 we have

$$g'(x) = \frac{2}{\pi} \frac{(x+1/2)}{\sqrt{\{1 - (x+1/2)^4\}}}$$

and the root was $\alpha = 0.2872$. It is clear that $g'(\alpha) \neq 0$ so that the scheme is first order. This is not unexpected since the scheme did converge fairly slowly. This example does not demonstrate the usefulness of knowing the order of a scheme because the root was calculated before the order of the scheme was determined.

Since the definition of order depends on the values of derivatives of $g(x)$ evaluated at a root $x = \alpha$ it may appear that the order of an iterative scheme will not be known until the root has been determined. If this were true our definition of order would be of no use. However it is usually a simple matter to use methods of algebra and calculus to determine the range of values of a given function over a specified interval. These techniques can be applied to the derivatives of $g(x)$ until the first non-vanishing derivative is found. In addition, higher order schemes can be derived analytically without any knowledge of the equation that is to be solved. The most common of these is the Newton–Raphson method, which is described in section 2.6. However, first we look at a very simple, foolproof method.

2.5 THE BISECTION METHOD

When a function $f(x)$ is continuous for $a < x < b$ and $f(a)$ and $f(b)$ have opposite signs the function $y = f(x)$ must cross the x-axis at some point within the interval $[a, b]$, i.e. the equation $f(x) = 0$ has a root in this interval. This root can be found by repeatedly halving the interval. Denoting the mid-point of $[a, b]$ by m the root will be in $[a, m]$ if $f(a)$ and $f(m)$ have opposite signs. Otherwise $f(m)$ and $f(b)$ will have opposite signs and the root will lie in $[m, b]$. Thus the half-interval containing the root can be determined by simply comparing the sign of $f(m)$ with the sign of $f(a)$ or $f(b)$. The process can then be repeated for the smaller interval containing the root. In this way a sequence of midpoints is generated that converges to the root.

This is a good, almost foolproof method. However, convergence is not as rapid as in some other methods.

Example 2.4 Carry out four steps of the bisection method for the root of

$$(2x + 1)^2 = 4 \cos \pi x$$

lying in the interval $[1/4, 1/3]$.
Let

$$f(x) = (2x + 1)^2 - 4 \cos \pi x$$

and let $x_1, x_2, \ldots$ denote the mid-points of successive intervals that contain the root α. Applying the bisection method, noting that $f(1/4) < 0$ and $f(1/3) > 0$, we write

$$0.2500(-) < \alpha < 0.3333(+),$$

where the sign in brackets indicates the sign of the function f at the ends of the interval. Proceeding in this manner

$$x_1 = (0.2500 + 0.3333)/2 = 0.2917, f(x_1) > 0$$
$$0.2500(-) < \alpha < 0.2917(+)$$

$$x_2 = (0.2500 + 0.2917)/2 = 0.2709, f(x_2) < 0$$
$$0.2709(-) < \alpha < 0.2917(+)$$

$$x_3 = (0.2709 + 0.2917)/2 = 0.2813, f(x_3) < 0$$
$$0.2813(-) < \alpha < 0.2917(+)$$

$$x_4 = (0.2813 + 0.2917)/2 = 0.2865, f(x_4) < 0$$
$$0.2865(-) < \alpha < 0.2917(+).$$

At this stage we only have two-decimal-place accuracy, i.e. $\alpha = 0.29$ to two decimal places. Further iterations would be necessary to increase the accuracy. Note that this root was also determined in Example 2.2 and was 0.2872 to four decimal places.

Algorithm 2.2. The bisection method

a & b are such that $f(a)^* f(b) < 0$; Tol is the tolerance.

```
read(a, b, Tol);
k := 0; fa := f(a)
REPEAT
    k := k + 1;
    mp := (a + b)/2;
    fmp := f(mp);
    IF fmp* fa > 0 THEN
```

```
      BEGIN
      a:= mp; fa:= fmp
      END
   ELSE b:= mp
   UNTIL (b − a) < Tol;
write(a, k).
```

2.6 THE NEWTON–RAPHSON METHOD FOR SOLVING $f(x) = 0$

This method can be derived by taking the tangent line to the curve $y = f(x)$ at the point $(x_n, f(x_n))$ corresponding to the current estimate, x_n, of the root. The intersection of this line with the x-axis gives the next estimate to the root, x_{n+1} (see Fig. 2.2).

The gradient of the curve $y = f(x)$ at the point $(x_n, f(x_n))$ is $f'(x_n)$ and the tangent line at this point has the form

$$y = f'(x_n)x + b.$$

Since this passes through $(x_n, f(x_n))$ we see that $b = f(x_n) - x_n f'(x_n)$ and therefore the tangent line is

$$y = f'(x_n)x + f(x_n) - x_n f'(x_n).$$

To determine where this line cuts the x-axis we set $y = 0$. Taking this point of intersection as the next estimate, x_{n+1}, to the root we have

$$0 = f'(x_n)x_{n+1} + f(x_n) - x_n f'(x_n)$$

and rearranging

$$x_{n+1} = x_n - \frac{f(x_n)}{f'(x_n)}. \tag{2.16}$$

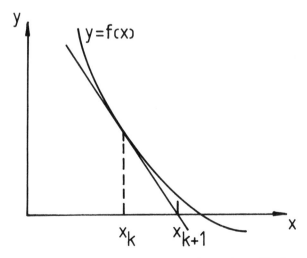

Fig. 2.2. Geometrical interpretation of the Newton–Raphson scheme. The intersection of the tangent to $y = f(x)$ at $(x_k, f(x_k))$ with the x-axis is taken as the next estimate, x_{k+1}, to the root.

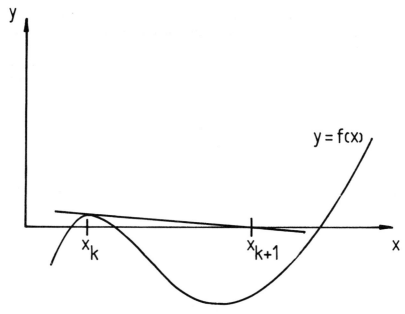

Fig. 2.3. The Newton–Raphson scheme can give difficulty when roots of $f(x) = 0$ and $f'(x) = 0$ are close together. This often arises when $f(x) = 0$ has two roots close together.

This is the *Newton–Raphson scheme*. This scheme has the form 'nest estimate = current estimate + correction term'. The correction term is $-f(x_n)/f'(x_n)$ and this must be small when x_n is close to the root if convergence is to be achieved. This will depend on the behaviour of $f'(x)$ near the root and, in particular, difficulty will be encountered when $f'(x)$ and $f(x)$ have roots close together (see Fig. 2.3).

Although the derivation of the Newton–Raphson method differs from the derivation of the simple methods in section 2.2, it is of the form $x_{n+1} = g(x_n)$ with

$$g(x) = x = \frac{f(x)}{f'(x)}. \tag{2.17}$$

Consequently the convergence criteria of section 2.3 can be applied to this function $g(x)$ and the order of the method can be examined. Differentiating (2.17) leads to

$$g'(x) = \frac{f(x)f''(x)}{(f'(x))^2} \tag{2.18}$$

and for convergence we require that

$$\left| \frac{f(x)f''(x)}{(f'(x))^2} \right| < 1 \tag{2.19}$$

for all x in some interval I containing the root. Since $f(\alpha) = 0$, the above condition is certainly satisfied at the root $x = \alpha$ provided that $f'(\alpha) \neq 0$. We consider the possibility of $f'(\alpha) = 0$ later. Then, provided that $g(x)$ is continuous, an interval I must exist in the neighbourhood of the root and over which (2.19) is satisfied. However, difficulty is sometimes encountered when the interval I is small because the initial guess must be taken from this interval. This usually arises when $f(x)$ and $f'(x)$ have roots close together since the correction term is inversely proportional to $f'(x)$. This feature is demonstrated graphically in Fig. 2.3. Note that the tangent line is almost parallel to the x-axis in the neighbourhood of the two roots that are close together. Consequently x_{n+1} can differ considerably from x_n unless an extremely accurate initial guess is obtained. In section 2.7 we present a method for finding an accurate initial guess when roots are close together.

Substituting $x = \alpha$ in equation (2.18) shows that $g'(\alpha) = 0$ provided that $f'(\alpha) \neq 0$. Consequently, *when $f'(\alpha) \neq 0$, the Newton–Raphson method is at least a second-order process*. However when the Newton–Raphson method is applied to an equation $f(x) = 0$ that has a multiple root $x = \alpha$ we find that $f'(\alpha) = 0$ and the Newton–Raphson scheme is only first order. To prove this, suppose that the equation $f(x) = 0$ has a root $x = \alpha$ of multiplicity m. Then we may write

$$f(x) = (x - \alpha)^m h(x)$$

which we substitute, together with

$$f'(x) = m(x - \alpha)^{m-1} h(x) + (x - \alpha)^m h'(x)$$

and

$$f''(x) = m(m - 1)(x - \alpha)^{m-2} h(x) + 2m(x - \alpha)^{m-1} h'(x) + (x - \alpha)^m h''(x),$$

into (2.18). Then after cancelling $(x - \alpha)^{2m-2}$ we have

$$g'(x) = \frac{h(x)\{m(m - 1)h(x) + 2m(x - \alpha)h'(x) + (x - \alpha)^2 h''(x)\}}{\{mh + (x - \alpha)^2 h'\}^2}$$

and substituting $x = \alpha$ this reduces to

$$g'(\alpha) = \frac{m - 1}{m}. \tag{2.20}$$

Thus

$$g'(\alpha) = 0 \quad m = 1$$
$$\neq 0 \quad m \neq 1,$$

i.e. the Newton–Raphson scheme is only first order for repeated roots ($m > 1$) but is at least second order for roots that are not repeated ($m = 1$). In addition $g'(\alpha)$ gets closer to unity as m increases and the convergence becomes less rapid. Exercise 20 shows modifications to the Newton–Raphson procedure that restore its order to two for the case $m > 1$.

Example 2.5 Use the Newton–Raphson procedure to determine the root of

$$(2x + 1)^2 = 4 \cos \pi x$$

lying in the interval $[1/4, 1/3]$.

To apply the Newton–Raphson scheme the equation must first be arranged into the form $f(x) = 0$. Thus we write

$$(2x + 1)^2 - 4 \cos \pi x = 0$$

and therefore

$$f(x) = (2x + 1)^2 - 4 \cos \pi x, \quad f'(x) = 4(2x + 1) + 4\pi \sin \pi x.$$

Substituting into equation (2.16) leads to the iterative scheme

$$x_{n+1} = x_n - \frac{\{(2x_n + 1)^2 - 4\cos \pi x_n\}}{\{4(2x_n + 1) + 4\pi \sin \pi x_n\}}.$$

Taking the mid-point of the interval as the initial guess we have $x_0 = 0.29167$ and the sequence of numbers generated by the iterative scheme is 0.28726, 0.28724, 0.28724. Thus to four decimal places the root is 0.2872. Note that, as expected, this second-order scheme converges far more rapidly than the simple iterative scheme and the bisection method that were used in Examples 2.2 and 2.3 respectively to solve the same equation.

Algorithm 2.3. The Newton–Raphson method, when $f'(x)$ not near to zero

x is an approximation to the root and Tol is the tolerance.

read(x, Tol);
$k := 0$;
REPEAT
   $k := k + 1$;
   CorrTerm := $f(x)/f'(x)$;
   $x := x - $ CorrTerm
   UNTIL abs(CorrTerm) < Tol;
write(x, k).

2.7 EQUATIONS WITH NEARLY EQUAL ROOTS

In the previous section we noted that, if two roots of the equation $f(x) = 0$ are close together, difficulty can be encountered when using the Newton–Raphson method to determine these roots. In this situation the convergence conditions will only be satisfied in a very small interval. Consequently an accurate initial guess is required, and here we show how to obtain such a guess.

If the equation $f(x) = 0$ has two nearly equal roots close to $x = \lambda$ (Fig. 2.3) then

$$f(\lambda) \simeq 0, \quad f'(\lambda) \simeq 0.$$

To obtain accurate initial guesses to these roots it is first necessary to find the root of $f'(x) = 0$ in the neighbourhood of λ. From the estimate to the root of $f'(x) = 0$ the two roots of $f(x) = 0$ can be found as follows.

Let μ be a root of $f'(x) = 0$ such that $\mu + \varepsilon$, where ε is small, is a root of $f(x) = 0$. Then, using a Taylor expansion (Chapter 1)

$$f(\mu + \varepsilon) = f(\mu) + \varepsilon f'(\mu) + \frac{\varepsilon^2}{2} f''(\mu) + \cdots$$

and since $f(\mu + \varepsilon) = 0$, $f'(\mu) = 0$, we have

$$f(\mu) + \frac{\varepsilon^2}{2} f''(\mu) \simeq 0.$$

Thus

$$\varepsilon^2 \simeq -2\frac{f(\mu)}{f''(\mu)}. \tag{2.21}$$

In summary, the procedure is first to calculate μ, the root of $f'(x) = 0$. Then ε can be calculated using (2.21) and $\mu + \varepsilon$, $\mu - \varepsilon$ can be used as initial estimates to the two roots of $f(x) = 0$. From these initial guesses the roots of $f(x) = 0$ can be determined accurately using the Newton–Raphson procedure.

Example 2.6 Determine the roots of the equation

$$x^4 - 1.99x^3 - 1.76x^2 + 5.22x - 2.23 = 0$$

that are close to $x = 1.5$.

Let

$$f(x) = x^4 - 1.99x^3 - 1.76x^2 + 5.22x - 2.23$$

and first find a root of $f'(x) = 0$. We have

$$f'(x) = 4x^3 - 5.97x^2 - 3.52x + 5.22$$

and if we relabel this as $h(x)$ we can find the root of $f'(x) = 0$ using the Newton–Raphson scheme

$$x_{k+1} = x_k - \frac{h(x_k)}{h'(x_k)} = x_k - \frac{f'(x_k)}{f''(x_k)}$$

with initial guess $x_0 = 1.5$. This leads to the sequence 1.498653501, 1.498649564 and then repeats the ten digits, i.e. the root μ of $f'(x) = 0$ has been found to at least nine significant figures. Next, using (2.21), we find that $\varepsilon^2 = 0.0049679399$ and consequently we obtain the initial guesses

$$\mu + \varepsilon = 1.569133179, \quad \mu - \varepsilon = 1.428165949$$

for the two roots of $f(x) = 0$ close to $x = 1.5$. Using these initial guesses with the Newton–Raphson scheme, $x_{k+1} = x_k - f(x_k)/f'(x_k)$, leads to the two sequences

$$
\begin{array}{ll}
1.565971998 & 1.424016937 \\
1.565887295 & 1.424112715 \\
1.565887234 & 1.424112766 \\
1.565887234 & 1.424112766
\end{array}
$$

giving convergence to at least nine significant figures after only four iterations, i.e. the roots are 1.42411277, 1.56588723.

Algorithm 2.4. The Newton–Raphson method for nearly equal roots

x is an approximation to the pair of nearly equal roots and Tol is the tolerance.

```
read(x, Tol);
y := x;
```

```
REPEAT
   CorrTerm := f'(y)/f"(y);
   y := y − CorrTerm
   UNTIL abs(CorrTerm) < Tol;
eps := sqrt(− 2*f(y)/f"(y));
Sign := 1;
FOR i := 1 TO 2 DO
   BEGIN
   k := 0;   y := x + Sign*eps;
   REPEAT
      k := k + 1;
      CorrTerm := f(y)/f'(y);
      y := y − CorrTerm
      UNTIL abs(CorrTerm) < Tol;
   Sign := − 1;
   write(y, k)
   END.
```

2.8 REAL ROOTS OF POLYNOMIAL EQUATIONS

The methods described earlier in this chapter can be used to solve polynomial equations. However, the method presented here is a more efficient and systematic method for extracting the roots of polynomial equations. An nth-degree polynomial equation has exactly n roots. As they are found, the corresponding factor can be removed from the polynomial by division. This avoids the possibility of the iterative scheme converging to the same root from different initial guesses.

The method we describe is an efficient application of the Newton–Raphson method to polynomial equations making use of some elementary algebraic ideas. Consequently it has the same order as the Newton–Raphson procedure which is two for distinct roots of the equation. We consider the polynomial equation

$$P_m(x) = 0 \tag{2.22}$$

where $P_m(x)$ is a polynomial of degree m taking the general form

$$P_m(x) = a_0 x^m + a_1 x^{m-1} + a_2 x^{m-2} + \cdots a_m. \tag{2.23}$$

The Newton–Raphson scheme for solving (2.22) is

$$x_{n+1} = x_n - \frac{P_m(x_n)}{P'_m(x_n)} \tag{2.24}$$

and we show how to evaluate the correction term, $-P_m(x_n)/P'_m(x_n)$, by carrying out two polynomial divisions. When dividing $P_m(x)$ by $(x - x_n)$ we will obtain a quotient of degree $m - 1$, which we denote by $Q_{m-1}(x)$, and a constant remainder r. Thus we write

$$P_m(x) = (x - x_n) Q_{m-1}(x) + r. \tag{2.25}$$

Substituting $x = x_n$ we have $r = P_m(x_n)$. This is the remainder theorem of elementary algebra and we have shown that the numerator in the correction term is the remainder when $P_m(x)$ is divided by $(x - x_n)$. We now show that the denominator in the correction term, $P'_m(x_n)$, is the remainder when $Q_{m-1}(x)$ is divided by $(x - x_n)$. To determine $P'_m(x_n)$ first we differentiate (2.25):

$$P'_m(x) = Q_{m-1}(x) + (x - x_n) Q'_{m-1}(x),$$

and therefore $P'_m(x_n) = Q_{m-1}(x_n)$. However, similar to equation (2.25) we may write

$$Q_{m-1}(x) = (x - x_n) S_{m-2}(x) + s, \qquad (2.26)$$

where $S_{m-2}(x)$ is the quotient, of degree $m - 2$, and s is the remainder when $Q_{m-1}(x)$ is divided by $(x - x_n)$. Substituting $x = x_n$ into equation (2.26) shows that $Q_{m-1}(x_n) = s$ and therefore $P'_m(x_n) = s$. Then substituting $P_m(x_n) = r$ and $P'_m(x_n) = s$ into (2.24) the Newton–Raphson scheme can be expressed in the form

$$x_{n+1} = x_n - \frac{r}{s}. \qquad (2.27)$$

Thus successive approximations to the roots of $P_m(x) = 0$ can be found by carrying out two divisions. These divisions can be carried out efficiently using *synthetic division* which we now describe. Denoting the coefficients of $Q_{m-1}(x)$ by $b_0, b_1, \ldots b_{m-1}$, equation (2.25) can be written

$$a_0 x^m + a_1 x^{m-1} + a_2 x^{m-2} + \cdots + a_m$$
$$= (x - x_n)(b_0 x^{m-1} + b_1 x^{m-2} + b_2 x^{m-3} + \cdots + b_{m-1}) + r.$$

We now compare coefficients of x^m, $x^{m-1}, \ldots$ in turn:

$$
\begin{aligned}
a_0 &= b_0 & b_0 &= a_0 \\
a_1 &= b_1 - x_n b_0 & b_1 &= a_1 + x_n b_0 \\
a_2 &= b_2 - x_n b_1 & b_2 &= a_2 + x_n b_1 \\
a_m &= r - x_n b_{m-1} & r &= a_m + x_n b_{m-1}.
\end{aligned}
$$

The left-hand set of equations is obtained by comparing coefficients directly. Since, in a given problem, the coefficients of $P_m(x)$, namely $a_0, a_1, \ldots, a_m$, will be known we rearrange these equations to make $b_0, b_1, \ldots, b_{m-1}, r$ the subjects of the equations. This leads to the right-hand set of equations. Note that each of $b_0, b_1, \ldots, b_{m-1}, r$ can be calculated in turn. Apart from the first and last equations, the equations for calculating these coefficients form a pattern. It is convenient to make the last equation fit into this pattern by relabelling the remainder r as b_m. Then the b's can be generated from the recurrence relation

$$b_0 = a_0, \quad b_k = a_k + x_n b_{k-1}, \quad 1 \leqslant k \leqslant m. \qquad (2.28)$$

Denoting the coefficients of $S_{m-2}(x)$ by $c_0, c_1, \ldots, c_{m-2}$ and relabelling s as c_{m-1} it can be shown in a similar manner that the c's can be generated from the recurrence relation

$$c_0 = b_0, \quad c_k = b_k + x_n c_{k-1}, \quad 1 \leqslant k \leqslant m - 1. \qquad (2.29)$$

Since we have relabelled the remainders r, s as b_m, c_{m-1} the Newton–Raphson scheme (2.27) can be written

$$x_{n+1} = x_n - \frac{b_m}{c_{m-1}}. \qquad (2.30)$$

This modification of the Newton–Raphson scheme is sometimes called the *Birge–Vieta* method. In its present form it can be used to determine distinct real roots of the polynomial equation $P_m(x) = 0$. We summarize the method in the following computational algorithm.

Algorithm 2.5. The Birge–Vieta method for polynomial equations

The degree of the polynomial $P(x)$ is m and its coefficients are $a[0], a[1], \ldots, a[m]$, with $a[1]$ the coefficient of the highest power. The initial approximation is x and the tolerance is Tol.

```
read(x,m,Tol);
FOR i:= 0 TO m DO read(a[i]);
k:= 0;
REPEAT
   k:= k + 1;
   b[0]:= a[0]; c[0]:= b[0];
   FOR i:= 1 TO m − 1 DO
      BEGIN
      b[i]:= a[i] + x*b[i − 1]; c[i]:= b[i] + x*c[i − 1]
      END;
   b[m]:= a[m] + x*b[m − 1];
   CorrTerm:= b[m]/c[m − 1];
   x:= x − CorrTerm
   UNTIL abs(CorrTerm) < Tol;
write(x, k).
```

This algorithm will provide an estimate to the root of $P_m(x) = 0$ in the neighbourhood of x_0. The accuracy of the estimate is determined by ε. For example, $\varepsilon = 0.00005$ will give four-decimal-place accuracy. Although the general form and calculation order of the algorithm should be preserved, a certain amount of flexibility is in the hands of the programmer. For example, although it may be necessary to put the coefficients a_i $(1 \leqslant i \leqslant m)$ into an array, the coefficients b_i $(1 \leqslant i \leqslant m)$ and c_i $(1 \leqslant i \leqslant m - 1)$ need not be stored in arrays.

Algorithm 2.5 is useful when the Birge–Vieta method is to be implemented in a program. However, if the method is to be used with a non-programmable hand calculator the following tabular form is a better approach:

a_0	a_1	a_2	$\cdots$	a_m
	$x_n b_0$	$x_n b_1$	$\cdots$	$x_n b_{m-1}$
b_0	b_1	b_2	$\cdots$	$b_{m-1}\, b_m$
	$x_n c_0$	$x_n c_1$	$\cdots$	$x_n c_{m-2}$
c_0	c_1	c_2		$c_{m-2}\, c_{m-1}$

The elements in the first row of the table are the coefficients of $P_m(x)$. All other rows will depend on the value of the current estimate x_n. The second and third rows are formed simultaneously working from left to right. The first element of the second row is zero. All other elements in the second row are equal to x_n times the element in the previous column of the third row. The elements in the third row are the sums of the corresponding elements in the first and second rows. It should be noted that the formation of the first three rows executes the recurrence relation (2.28). In a similar manner the last three rows execute the recurrence relation (2.29). We have underlined the numbers b_m and c_{m-1} that are involved directly in the Newton–Raphson formula (2.30). Note that a new table is required at every stage of an iteration until convergence is achieved.

Example 2.7 Use the Birge–Vieta method to find the root of

$$8x^3 - 20x^2 - 26x + 33 = 0$$

close to 0.8.

Taking the initial guess $x_0 = 0.8$ we set out the table as indicated above:

8.00000	− 20.00000	− 26.00000	33.00000
	6.40000	− 10.80000	− 29.50400
8.00000	− 13.60000	− 36.88000	3.49600
	6.40000	− 5.76000	
8.00000	− 7.20000	− 42.64000	

The first row is formed from the coefficients of the polynomial. Then 8.00000 is entered in the third row followed by $x_0 \times 8.00000 = 6.40000$ in the second row. Next the second element of the third row is calculated by adding the two numbers above it, yielding − 13.60000. This number is multiplied by x_0 to give − 10.80000 in the next vacant position of the second row. Proceeding in this manner the second and third rows are calculated. The fourth and fifth rows are calculated from the first row. Note that the whole of this table is calculated from the coefficients in the polynomial equation and the initial guess x_0. Having completed the table we calculate x_1 from the Birge–Vieta formula (2.29):

$$x_1 = x_0 - \frac{b_3}{c_2} = 0.80000 - \frac{-3.49600}{(-42.64000)}$$

$$= 0.88199.$$

We will aim for four-decimal-place accuracy. Consequently all calculations will be carried out to five decimal places. Since x_0 and x_1 do not agree to four decimal places we construct a table for the next stage of the iteration. This table is formed in the same way as the above table but $x_1 = 0.88199$ is used instead of $x_0 = 0.80000$:

8.00000	− 20.00000	− 26.00000	33.00000
	7.05592	− 11.41655	− 33.00102
8.00000	− 12.94408	− 37.41655	− 0.00102
	7.05592	− 5.19330	
8.00000	− 5.88816	− 42.60984	

Using equation (2.30) we calculate the new estimate

$$x_2 = x_1 - \frac{b_3}{c_2} = 0.88199 - \frac{(-0.00102)}{(-42.60984)}$$

$$= 0.88199.$$

We see that x_2 and x_1 agree, when rounded to four decimal places. Thus to four decimal places, the root close to $x = 0.8$ is $\alpha = 0.8820$.

Recall that the third row contains the coefficients of the reduced polynomial and the remainder, which we have made sufficiently close to zero to obtain four decimal place accuracy, i.e. we have

$$8x^3 - 20x^2 - 26x + 33 \simeq (x - 0.8820)(8x^2 - 12.9441x - 37.4166).$$

Thus the other roots can be found by solving the quadratic equation

$$8x^2 - 12.9441x^2 - 27.4166 = 0.$$

However, because of the calculations that we have carried out to reach this stage the roots of this quadratic should only be regarded as approximations to the roots of the original cubic equation, i.e. they may contain round-off error. Consequently the roots of the quadratic equation should be used as initial guesses for the Birge–Vieta scheme which should then be applied to the original cubic equation. The reader may like to carry out these calculations. The other roots are -1.5000 and 3.1180.

EXERCISES 2

1. Initial guesses to the three roots of

$$x^3 - x = 0.1$$

can be obtained at a glance. What are they? Rearrange the equation algebraically to obtain recurrence relations of the form $x_{k+1} = g(x_k)$. Hence estimate the roots to four decimal places.

2. Which of the iterative schemes

$$x_{k+1} = \tan x_k \quad \text{or} \quad x_{k+1} = \tan^{-1} x_k$$

is suitable for solving the equation $x = \tan x$? Hence find the smallest positive root of $x = \tan x$.

3. Show that the equation

$$xe^x = 1$$

has a root between 0.5 and 0.6. Which of the iterative schemes

$$x_{k+1} = -\log x_k \quad \text{or} \quad x_{k+1} = \exp(-x_k)$$

is suitable for calculating this root? Hence determine this root to three decimal places.

4. Show that the equation

$$x^3 - 12x + 3 = 0$$

has one root in each of the intervals $(-4, -3), (0, 1)$ and (3.4). Show that the scheme

$$x_{k+1} = \tfrac{1}{12}(x_k^3 + 3)$$

is suitable for finding one of these roots but unsuitable for the other two roots. Determine this root to four decimal places.

   Determine the other two roots approximately by factorizing the cubic. Devise iterative schemes, by algebraic rearrangement of the original equation, that will converge to these two roots and calculate these roots to four decimal places.

5. A possible scheme for finding the roots of the equation

$$2x^3 - 2x^2 - 3x + 1 = 0$$

is

$$x_{k+1} = \tfrac{1}{3}(2x_k^3 - 2x_k^2 + 1).$$

Show that, with a suitable initial guess, this scheme will converge to the root inside the interval (a, b), where a, b are the roots of the quadratic

$$6x^2 - 4x - 3 = 0.$$

Determine this root to three decimal places.

   Determine the other two roots approximately by factorizing the cubic. By rearranging the cubic equation, find schemes that will converge to these two roots. Use these schemes to check or improve the accuracy to three decimal places.

6. Determine the order of the following iterative processes for finding the roots of the equation is Exercise 4:

   (i) $x_{k+1} = \tfrac{1}{12}(x_k^3 + 3)$, root in $(0, 1)$,
   (ii) $x_{k+1} = (12x_k - 3)^{1/3}$, root in $(3, 4)$.

7. Show that the iterative scheme

$$x_{k+1} = \tfrac{1}{3}(4x_k - ax_k^4)$$

is a second-order process for the calculation of the reciprocal of the cube root of a. Use this scheme to estimate $3^{-1/3}$ to four decimal places.

8. Show that the iterative scheme

$$x_{k+1} = x_k(a^2x_k^2 - 4ax_k + 5)$$

is a second-order process for the calculation of $1/a$. Show that the scheme will not converge to the other two roots of the corresponding algebraic equation.

9. Show that the iterative scheme

$$x_{k+1} = \frac{1}{cn}[(n+1)cx_k - x_k^{n+1}]$$

is a second-order scheme for calculating the nth root of c. Use this scheme to calculate the fifth root of 2 to six decimal places.

10. Show that the $1/\sqrt{c}$ is a root of the converged equation corresponding to the

iterative scheme

$$x_{k+1} = \frac{1}{s}\{cx_k^3 - 1(1+s)x_k\},$$

where s is a constant. Show that the scheme will only converge if s is negative. Determine the value of s that will make the scheme have second order.

11. A root of the equation $f(x) = 0$ is denoted by α, and $x_{n+1} = g(x_n)$ is an iterative scheme that converges to α. By writing equation (2.14) in the form

$$\varepsilon_{n+1} = \varepsilon_n(g'(\alpha) + E_n)$$

where

$$E_n = \sum_{k=2}^{\infty} \frac{\varepsilon_n^{k-1}}{k!} g^{(k)}(\alpha)$$

and

$$\varepsilon_n = x_n - \alpha$$

show that

$$\frac{\varepsilon_{n+1}}{\varepsilon_n} - \frac{\varepsilon_{n+2}}{\varepsilon_{n+1}} = E_n - E_{n+1}.$$

Deduce that, for sufficiently large n,

$$\alpha \simeq \frac{x_n x_{n+2} - x_{n+1}^2}{x_{n+2} - 2x_{n+1} + x_n}.$$

This is called Aitken's δ^2-process for improving first-order root approximations, i.e. the right-hand side gives an improved approximation in terms of the last three successive approximations. With the notation of Chapter 3 it can also be written in the form

$$\alpha \simeq x_n - \frac{(\delta x_{n+1/2})^2}{\delta^2 x_{n+1}}$$

which explains the term 'δ^2-process'.

12. Starting with $x_0 = 0.5$, carry out a few iterations with the scheme

$$x_{k+1} = 1 - \sin x_k$$

to see how slowly it converges to a root of

$$\sin x + x - 1 = 0.$$

Also apply the iterative scheme using Aitken's δ^2-process after every three successive approximations.

13. Starting with the interval $(1, 2)$ use the bisection method to calculate the cube root of 3 to two decimal places (Hint: construct a simple equation with a root equal to $\sqrt[3]{3}$).

14. Locate an interval containing the root of $xe^x = 1$ from a rough sketch. Use the bisection method to determine this root to two decimal places.

15. An interval (a, b) is located that contains the root α of an equation $f(x) = 0$. It is

clear that n applications of the bisection method will ensure that

$$|x_n - \alpha| \leqslant (b - a)/2^n.$$

Show that the accuracy $|x_n - \alpha| < \varepsilon$ will be achieved when

$$n > (\ln [(b - a)/\varepsilon])/\ln 2.$$

16. Use the Newton–Raphson method to calculate $\sqrt{7}$ to four decimal places.

17. Show that the equation

$$x^3 - 6.37x^2 + 6.48x + 7.11 = 0$$

has a root in the interval $(4, 5)$ and use the Newton–Raphson method to determine this root to four decimal places.

18. Show that the equation

$$x^3 - 8.3841x^2 + 13.3046x + 2.3841 = 0$$

has a root near $x = 6$ and calculate this root to four decimal places using the Newton–Raphson method.

19. Estimate the root of the equation

$$x + \sinh x \cosh x = 1$$

to six decimal places.

20. Equation (2.20) shows that the Newton–Raphson method is only a first-order method for calculating a root of multiplicity $m > 1$ of the equation $f(x) = 0$. Show that the modified schemes

(i) $x_{k+1} = x_k - \dfrac{m f(x_k)}{f'(x_k)}$

(ii) $x_{k+1} = x_k - \dfrac{h(x_k)}{h'(x_k)}$

where $h(x) = f(x)/f'(x)$ are both of order two or greater.

21. The equation

$$16x^4 - 32x^3 + 8x^2 + 8x + 1 = 0$$

has a repeated root close to $x = 1$. Use one of the methods in Exercise 20 to determine this root to four decimal places.

22. The equation

$$5x^3 - 10x^2 + 6.232x - 1.232 = 0$$

has two roots close to $x = 0.5$. Calculate these roots to six decimal places.

23. Two roots of the equation

$$x^4 - 3.9080x^3 + 4.8083x^2 - 3.9080x + 3.8083 = 0$$

are close to $x - 2$. Calculate these roots to four decimal places.

24. Use the Birge–Vieta method to determine the roots of the equations in Exercises 17 and 18 to four decimal places.

3

Interpolation

The instrumentation on a satellite might be programmed to measure atmospheric densities at longitudes of 0°, 10°, 20°, 30°,... as the satellite orbits the Earth. Having obtained these data we may want to estimate densities at intermediate longitudes. To achieve this, the data could be plotted—density against longitude—and adjacent points joined with straight lines (Fig. 3.1). The values on the straight lines can then be used as estimates of the densities at intermediate longitudes. This process of using straight lines to estimate intermediate values is called *linear interpolation*.

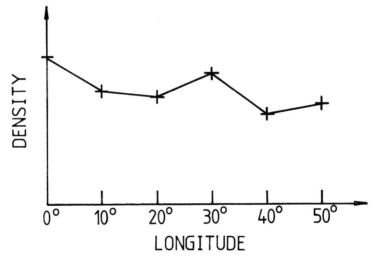

Fig. 3.1. When densities are known at 10° intervals in longitude they can be estimated at intermediate longitudes by joining adjacent points with straight lines. This is linear interpolation, the simplest form of interpolation.

More generally we consider the following problem. The analytic form of a function $y = f(x)$ is not known. The only information we have is that the function takes certain numerical values f_i when $x = x_i$ for $i = 0(1)n$ and we wish to estimate $f(x)$ for $x_i < x < x_{i+1}$, e.g. we may wish to estimate $f(0.27)$ given that $x, f(x)$ take the values shown in Table 3.1. The data points could be plotted in the xy-plane and adjacent points connected by straight lines. However this may not be sufficiently accurate and we therefore consider how to pass quadratics, cubics and even higher degree polynomials through a given set of points. An estimate of $f(x)$ for $x_i < x < x_{i+1}$ can then be obtained from one of these polynomials. This method of obtaining estimates of $f(x)$ from numerical data by fitting polynomials to the data is called *interpolation*. These same polynomials can be used to estimate $f(x)$ when x lies outside the range $[x_0, x_n]$. This is called *extrapolation*.

Recall from elementary mathematics that three data points are required to determine a, b, c for a quadratic $y = ax^2 + bx + c$, four data points are required to determine the four coefficients in a cubic and for a polynomial of degree n we require $(n + 1)$ data points to determine the $n + 1$ coefficients. Thus to find the quadratic passing through three successive data points (x_j, f_j), (x_{j+1}, f_{j+1}), (x_{j+2}, f_{j+2}), for example, we could substitute the three pairs of numerical values for x and y into the quadratic. This would lead to three simultaneous equations in a, b, c that could easily be solved. The resulting quadratic could be used to estimate $f(x)$ for any value of x between x_j and x_{j+2}. Although this approach appears simple there are two distinct disadvantages. The first is that we would have no idea of the accuracy of the estimate from the quadratic. The second is that with higher degree polynomials, larger systems of linear equations have to be solved to determine the coefficients. Because of this we use a different approach that overcomes both of these problems.

Consider the problem of estimating $f(0.26)$ when the only information available is that $f(x)$ takes the values 0.0000, 0.4054, 0.8441, 1.3547, 1.9878, 2.8134 for $x = 0.0(0.2)1.0$. The first step is to construct a *finite difference table* from the given data (Table 3.1). The formation of the table is self-evident, e.g. those marked with an asterisk are $0.5106 = 1.3547 - 0.8441$, $0.1225 = 0.6331 - 0.5106$. The difference columns are constructed from left to right. The first difference column is obtained by forming the differences of adjacent function values, i.e. calculating $f(x_{i+1}) - f(x_i)$ where x_i, x_{i+1} are

Table 3.1

x	$f(x)$	First differences	Second differences	Third differences
0.0	0.0000			
		0.4054		
0.2	0.4054		0.0333	
		0.4387		0.0386
0.4	0.8441		0.0719	
		0.5106*		0.0506
0.6	1.3547		0.1225*	
		0.6331		0.0700
0.8	1.9878		0.1925	
		0.8256		
1.0	2.8134			

adjacent x values with $x_{i+1} > x_i$. The remaining difference columns are calculated in a similar manner, i.e. by forming the differences of the adjacent numbers in the column immediately to the left. The values that appear in the difference columns are called *finite differences*. The number of difference columns that are needed depends on the accuracy required for the value of $f(x)$ that is to be interpolated. This cannot be decided until we have the interpolation polynomials available.

Throughout we will refer to a tabulated function. By this we mean a function that is only specified numerically and can be put into tabular form. We are going to derive formulae that involve differences of a tabular function such as those that appear in Table 3.1. These formulae are infinite series and can be used to estimate $f(x)$ for a non-tabular value of x. The first few terms of these series normally give sufficient accuracy and represent a polynomial passing through an adjacent set of data points. A polynomial passing through a given set of data points is called an interpolation polynomial.

3.1 FINITE DIFFERENCES AND DIFFERENCE OPERATORS

It will become evident that finite differences do not allow for data that are unequally spaced with respect to x. For such data, divided differences are needed (section 3.3). Thus in this section we restrict attention to data for which $x_{j+1} - x_j = h$, where the spacing h takes the same value for every pair of x-values x_j, x_{j+1}. This restriction applies whenever finite differences are used.

Finite difference notation

Since we are going to construct formulae involving finite differences it is necessary to introduce a notation to represent them. Labelling the data $(x_0, f_0), (x_1, f_1), \ldots$, where $f_j = f(x_j)$, the first three columns take the form

x	$f(x)$	First differences
x_0	f_0	
		$f_1 - f_0$
x_1	f_1	
		$f_2 - f_1$
x_2	f_2	
		$f_3 - f_2$
x_3	f_3	
		$f_4 - f_3$
x_4	f_4	

A first difference of f is represented by one of three notations (in this application each of Δ, ∇, δ are pronounced 'delta')

$$\Delta f, \nabla f, \delta f$$

and a suffix must be attached to indicate which first difference we are describing. For the difference $f_{j+1} - f_j$ the obvious choices for the suffix are j, $j+1$ or $j+1/2$. The convention is to associate each of these suffices with the three notations, i.e. $f_{j+1} - f_j$ can be denoted by $\Delta f_j, \Delta f_{j+1}$, or $\delta f_{j+1/2}$. Thus we have the *forward difference notation*

$$\Delta f_j = f_{j+1} - f_j, \tag{3.1}$$

the *backward difference notation*

$$\nabla f_{j+1} = f_{j+1} - f_j$$

or, replacing j by $j-1$,

$$\nabla f_j = f_j - f_{j-1} \tag{3.2}$$

and the central difference notation

$$\delta f_{j+1/2} = f_{j+1} - f_j. \tag{3.3}$$

We now extend each notation in the obvious way through the difference columns of a table (Tables 3.2–3.4).

For later use we draw attention to the *difference bands that have the same suffix*. An example of each difference band is marked in Tables 3.2–3.4 and we see that

(i) Δ-bands are diagonals with negative slope,

Table 3.2 Forward difference notation

x	$f(x)$	First differences	Second differences	Third differences
x_0	f_0			
		$f_1 - f_0 = \Delta f_0$		
x_1	f_1		$\Delta f_1 - \Delta f_0 = \Delta^2 f_0$	
		$f_2 - f_1 = \Delta f_1$		$\Delta^2 f_1 - \Delta^2 f_0 = \Delta^3 f_0$
x_2	f_2		$\Delta f_2 - \Delta f_1 = \Delta^2 f_1$	
		$f_3 - f_2 = \Delta f_2$		$\Delta^2 f_2 - \Delta^2 f_1 = \Delta^3 f_1$
x_3	f_3		$\Delta f_3 - \Delta f_2 = \Delta^2 f_2$	
		$f_4 - f_3 = \Delta f_3$		
x_4	f_4			

Table 3.3 Backward difference notation

x	$f(x)$	First differences	Second differences	Third differences
x_0	f_0			
		$f_1 - f_0 = \nabla f_1$		
x_1	f_1		$\nabla f_2 - \nabla f_1 = \nabla^2 f_2$	
		$f_2 - f_1 = \nabla f_2$		$\nabla^2 f_3 - \nabla^2 f_2 = \nabla^3 f_3$
x_2	f_2		$\nabla f_3 - \nabla f_2 = \nabla^2 f_3$	
		$f_3 - f_2 = \nabla f_3$		$\nabla^2 f_4 - \nabla^2 f_3 = \nabla^3 f_4$
x_3	f_3		$\nabla f_4 - \nabla f_3 = \nabla^2 f_4$	
		$f_4 - f_3 = \nabla f_4$		
x_4	f_4			

Table 3.4 Central difference notation

x	$f(x)$	First differences	Second differences	Third differences
x_0	f_0			
		$f_1 - f_0 = \delta f_{1/2}$		
x_1	f_1		$\delta f_{3/2} - \delta f_{1/2} = \delta^2 f_1$	
		$f_2 - f_1 = \delta f_{3/2}$		$\delta^2 f_2 - \delta^2 f_1 = \delta^3 f_{3/2}$
x_2	f_2		$\delta f_{5/2} - \delta f_{3/2} = \delta^2 f_2$	
		$f_3 - f_2 = \delta f_{5/2}$		$\delta^2 f_3 - \delta^2 f_2 = \delta^3 f_{5/2}$
x_3	f_3		$\delta f_{7/2} - \delta f_{5/2} = \delta^2 f_3$	
		$f_4 - f_3 = \delta f_{7/2}$		
x_4	f_4			

(ii) ∇-bands are diagonals with positive slope,
(iii) δ-bands are horizontal bands.

These bands are particularly useful when using interpolation formulae which involve differences with the same suffix. To extract a specific finite difference from a table

(i) start in the function value column at the value indicated by the suffix,
(ii) move along the appropriate delta-band to the required difference column.

As an example, take $x_0 = 0.4$ in Table 3.1 and therefore $x_1 = 0.6$, $x_{-1} = 0.2$, etc. To find the value of $\nabla^3 f_2$ in this table we start at $f_2 = f(0.8) = 1.9878$ and then move along a ∇-band until we reach the third difference column, giving $\nabla^3 f_2 = 0.0506$. In a similar manner it is easily checked that $\delta^2 f_1 = 0.0333$, $\Delta^3 f_2 = 0.0386$ and $\delta^3 f_{3/2} = 0.0700$ for Table 3.1 with $x_0 = 0.4$.

Finite difference operators

Associated with the notation for finite differences we define the forward, backward and central difference operators. These are operators on functions, just like the differential operator $D \equiv d/dx$, i.e. $Dx^2 = 2x$. Recall that

$$\Delta f(x_j) = \Delta f_j = f_{j+1} - f_j$$
$$= f(x_{j+1}) - f(x_j)$$
$$= f(x_j + h) - f(x_j)$$

and consistent with this notation we define the *forward difference operator* Δ by

$$\Delta f(x) = f(x + h) - f(x), \tag{3.4}$$

where $f(x)$ is any function of x. This generalizes the notation, since putting $x = x_j$ in (3.4) gives equation (3.1).

Example 3.1 Determine $\Delta f(x)$ for $f(x) = x^3 + 2x^2 + 1$.
We have

$$f(x + h) = (x + h)^3 + 2(x + h)^2 + 1$$
$$= x^3 + 3x^2 h + 3xh^2 + h^3 + 2(x^2 + 2xh + h^2) + 1$$

and therefore

$$\Delta f(x) = f(x+h) - f(x) = 3hx^2 + h(3h+4)x + h^2(h+2).$$

Note, in particular, that $f(x)$ was cubic and $\Delta f(x)$ turned out to be quadratic. Later we show that, if $p_m(x)$ is a polynomial of degree m, $\Delta p_m(x)$ is a polynomial of degree $m-1$. Exercises such as Example 3.1 are rarely necessary in practice but are helpful to gain familiarity with the difference operators.

Example 3.2 Show that

$$\Delta^3 f_k = f_{k+3} - 3f_{k+2} + 3f_{k+1} - f_k.$$

By repeated application of the difference operator Δ and using (3.4) with $x = x_k$, x_{k+1}, x_{k+2} as appropriate we have

$$\Delta f_k = f_{k+1} - f_k$$
$$\Delta^2 f_k = \Delta(\Delta f_k) = \Delta f_{k+1} - \Delta f_k$$
$$= (f_{k+2} - f_{k+1}) - (f_{k+1} - f_k)$$
$$= f_{k+2} - 2f_{k+1} + f_k$$
$$\Delta^3 f_k = \Delta(\Delta^2 f_k) = \Delta f_{k+2} - 2\Delta f_{k+1} + \Delta f_k$$
$$= (f_{k+3} - f_{k+2}) - 2(f_{k+2} - f_{k+1}) + (f_{k+1} - f_k)$$
$$= f_{k+3} - 3f_{k+2} + 3f_{k+1} - f_k.$$

A similar result can be obtained from Table 3.2 rather than equation (3.4) by starting with $\Delta^4 f_0 = \Delta^3 f_1 - \Delta^2 f_0$, substituting $\Delta^2 f_1 = \Delta f_2 - \Delta f_1$ and $\Delta^2 f_0 = \Delta f_1 - \Delta f_0$ and then substituting for the first differences.

Similarly, recalling that

$$\nabla f(x_j) = \nabla f_j = f_j - f_{j-1}$$
$$= f(x_j) - f(x_{j-1})$$
$$= f(x_j) - f(x_j - h)$$

we define the *backward difference operator* ∇ by

$$\nabla f(x) = f(x) - f(x-h) \tag{3.5}$$

and, similar to the above example, it is easily shown that

$$\nabla^4 f_k = f_k - 4f_{k-1} + 6f_{k-2} - 4f_{k-3} + f_{k-4}.$$

Once it is realized that these expressions have the binomial coefficients with alternating signs they can be written down immediately. Later we obtain more general expressions for $\Delta^r f_k$ and $\nabla^r f_k$ which involve binomial coefficients. Although such expressions are of little use for interpolation they are needed later for numerical integration.

Finally we recall that

$$\delta f(x_j + h/2) = \delta f_{j+1/2} = f_{j+1} - f_j$$
$$= f(x_{j+1}) - f(x_j)$$
$$= f(x_j + h) - f(x_j)$$

and consequently we define the *central difference operator* δ by

$$\delta f(x + h/2) = f(x+h) - f(x)$$

or, replacing x by $x - h/2$,

$$\delta f(x) = f(x + h/2) - f(x - h/2). \tag{3.6}$$

Example 3.3 Determine $\delta f(x)$ when $f(x) = a$ and a is a constant.
  Since $f(x) = a$, where a is a constant, we have $f(x + h/2) = a$ and $f(x - h/2) = a$ (consider the graph of $f(x) = a$) and therefore

$$\delta f(x) = a - a = 0.$$

Similarly $\Delta f(x) = 0$ and $\nabla f(x) = 0$ when $f(x) = a$.
  In view of this example it is particularly important to observe that $\Delta f(x) \neq f(x)\Delta$, for example. The only possible interpretation for $f(x)\Delta$ is $f(x)\Delta(1) = 0$ and $\Delta f(x)$ is not equal to zero for all functions $f(x)$. Thus it is essential to preserve the order

(operator)(function).

The shift operator E

When using finite differences or finite difference operators we frequently encounter quantities such as $f_{j+1} = f(x_{j+1}) = f(x_j + h)$ or $f(x + h)$. Because of this it is useful to define the *shift operator* E by

$$Ef(x) = f(x + h), \tag{3.7}$$

It follows that $Ef(x_j) = f(x_j + h) = f(x_{j+1})$ or

$$Ef_j = f_{j+1}. \tag{3.8}$$

By repeatedly operating E on each side of (3.7) we have

$$E^2 f(x) = E(f(x + h)) = f(x + 2h)$$
$$E^3 f(x) = Ef(x + 2h) = f(x + 3h)$$
$$\vdots$$
$$E^r f(x) = f(x + rh), \tag{3.9}$$

where r is a positive integer. We now attach a meaning to the inverse E-operator E^{-1} which is such that $EE^{-1}f(x) = E^{-1}Ef(x) = f(x)$. Operating E^{-1} on (3.7) gives

$$E^{-1}f(x + h) = E^{-1}Ef(x) = f(x)$$

and, since this must be true for all values of x, we can replace x by $x - h$:

$$E^{-1}f(x) = f(x - h), \tag{3.10}$$

i.e. E^{-1} represents a backward shift and similar to (3.8) we have

$$E^{-1}f_j = f_{j-1}. \tag{3.11}$$

By repeatedly operating E^{-1} on (3.10) we obtain, similar to (3.9),

$$E^{-r}f(x) = f(x - rh) \tag{3.12}$$

and, to summarize, equation (3.9) now holds for all negative and positive integer values of r.

The averaging operator μ

Some of the interpolation formulae that we derive later involve quantities such as

$$\tfrac{1}{2}(f_0 + f_1), \tfrac{1}{2}(\delta f_{-1/2} + \delta f_{1/2}), \tfrac{1}{2}(\delta^3 f_{-1/2} + \delta^3 f_{1/2}).$$

These quantities arise when taking the average of two formulae and we find it useful to introduce a more concise notation to represent them. Given that we are going to use the letter μ to indicate the averaging process the above quantities could be denoted by μf, $\mu \delta f$ and $\mu \delta^3 f$ with a suitable suffix attached. The most sensible suffix to choose is the average of the two suffices on the function values or finite differences. Thus we denote each of the above quantities by $\mu f_{1/2}$, $\mu \delta f_0$ and $\mu \delta^3 f_0$. Consistent with this notation we define the averaging operator μ by

$$\mu f(x) = \tfrac{1}{2}\{f(x + h/2) + f(x - h/2)\} \tag{3.13}$$

for any function $f(x)$. Then setting $x = x_j + h/2$, for example, gives

$$\mu f(x_j + h/2) = \tfrac{1}{2}\{f(x_j + h) + f(x_j)\}$$

or

$$\mu f_{j+1/2} = \tfrac{1}{2}\{f_{j+1} + f_j\}. \tag{3.14}$$

Similarly, replacing $f(x)$ by $\delta^3 f(x)$, for example, and setting $x = x_j$ in (3.13) lead to

$$\mu \delta^3 f_j = \tfrac{1}{2}\{\delta^3 f_{j+1/2} + \delta^3 f_{j-1/2}\}. \tag{3.15}$$

Relations between the operators

Since we have introduced three notations for a given finite difference it is to be expected that we can relate the associated operators Δ, ∇ and δ. We have also introduced the shift operator E and averaging operator μ. Many equations can be derived that relate these operators. In particular, the shift operator can be regarded as the basic operator from which the others can be defined. Some of these equations are

$$\Delta = E - 1 \tag{3.16}$$
$$\nabla = 1 - E^{-1} \tag{3.17}$$
$$\delta = E^{1/2} - E^{-1/2} \tag{3.18}$$
$$\mu = \tfrac{1}{2}(E^{1/2} + E^{-1/2}). \tag{3.19}$$

Their proofs follow easily from the definitions of the operators. For example, by definition

$$\Delta f_j = f_{j+1} - f_j$$

and

$$E f_j = f_{j+1}.$$

Thus

$$\Delta f_j = E f_{j+1} - f_j$$
$$= (E - 1)f_j$$

showing that the operators Δ and $E - 1$ are equivalent and proving (3.16). The other proofs are similar (Exercise 4). Later we make use of the operator equations

$$E = 1 + \Delta \tag{3.20}$$

$$E = (1 - \nabla)^{-1} \tag{3.21}$$

$$E = 1 + \frac{\delta^2}{2} + \delta\sqrt{(1 + \delta^2/4)} \tag{3.22}$$

which follow from equations (3.16), (3.17), (3.18) (Exercise 4).
From equation (3.16)

$$\Delta^k = E^k(1 - 1/E)^k$$

$$= E^k\left\{1 + k\left(-\frac{1}{E}\right) + \frac{k(k-1)}{2!}\left(-\frac{1}{E}\right)^2 + \cdots + \left(-\frac{1}{E}\right)^k\right\}$$

or

$$\Delta^k = \sum_{r=0}^{k} \frac{(-1)^r k!}{r!(k-r)!} E^{k-r}$$

and therefore, operating on f_j, we obtain

$$\Delta^k f_j = \sum_{r=0}^{k} \frac{(-1)^r k!}{r!(k-r)!} f_{j+k-r}, \tag{3.23}$$

i.e. $\Delta^k f_j$ can be expressed as a linear combination of $f_j, f_{j+1}, \ldots, f_{j+k}$ and the coefficients are the binomial coefficients with alternating signs. Similar expressions, also involving the binomial coefficients with alternating signs, can be derived for $\nabla^k f_j$ and $\delta^k f_j$. Thus equations (3.17) and (3.18) lead to

$$\nabla^k f_j = \sum_{r=0}^{k} \frac{(-1)^r k!}{r!(k-r)!} f_{j-r} \tag{3.24}$$

and

$$\delta^k f_j = \sum_{r=0}^{k} \frac{(-1)^r k!}{r!(k-r)!} f_{j+k/2-r}, \tag{3.25}$$

showing that $\nabla^k f_j$ can be expressed as a linear combination of $f_j, f_{j-1}, \ldots, f_{j-k}$ and $\delta^k f_j$ can be expressed as a linear combination of $f_{j-k/2}, \ldots, f_j, \ldots, f_{j+k/2}$. When applying equation (3.25) to a given set of data (x_j, f_j), $j = 0(1)n$, the f-suffix must always be an integer, i.e. $j + k/2$ must be an integer. Thus when $\delta^k f_j$ is an even-order central difference (k even) the suffix j must be an integer and when $\delta^k f_j$ is an odd-order central difference (k odd) the suffix j must be a mid-point suffix, i.e. integer $+ 1/2$. This is consistent with the notation in Table 3.4.

The importance of the shift operator in interpolation

In interpolation we start with a set of data (x_j, f_j), $j = 0(1)n$, and we consider the problem of estimating $f(x)$ when x lies between any pair of $x_0, x_1, \ldots, x_n$. In many problems these x-values are equally spaced. Denoting the spacing by h we represent this fact by writing

$$x_{j+1} = x_j + h. \tag{3.26}$$

With equally spaced data it is more convenient to work in terms of a dimensionless variable r measured in units of h and defined by

$$x = x_j + rh, \quad -1/2 \leqslant r \leqslant 1/2, \tag{3.27}$$

where x_j is the closest of $x_0, x_1, \ldots, x_n$ to x. The inequality for r expresses the fact that x lies within a distance $h/2$ from x_j. Consequently the value that we wish to interpolate is

$$f(x) = f(x_j + rh), \quad -1/2 \leqslant r \leqslant 1/2,$$

and to get to $x = x_j + rh$ we have to *shift* a fraction of step length, the fraction being determined by r, away from the closest point x_j. This leads to a logical extension of the definition of the shift operator (see equations (3.9), (3.12)), namely that

$$E^r f(x) = f(x + rh) \tag{3.28}$$

for *any value of* r. In particular, the value that we want to interpolate can now be represented by $E^r f_j$ or $E^r f(x_j)$ since

$$f(x) = f(x_j + rh) = E^r f(x_j) = E^r f_j, \quad -1/2 \leqslant r \leqslant 1/2. \tag{3.29}$$

Thus we can use the shift operator to shift part way between two tabular points and this can be related to the problem of estimating $f(x)$ when x lies between any two tabular points. Equation (3.29) gives us the notation $E^r f_j$ for the value $f(x) = f(x_j + rh)$ that we want to interpolate. In section 3.3 we express $f(x) = E^r f_j$ in terms of forward and backward differences, i.e. the numbers that appear in finite difference tables. However, first we investigate whether or not it is reasonable to approximate a function by a polynomial.

Differences of a polynomial; approximation of a function by a polynomial

Before fitting polynomials to a given set of data we need to examine whether or not a polynomial can give a reasonable representation of the data. To approach this problem we first consider the process of differencing a polynomial. We consider a general polynomial of degree m:

$$p_m(x) = a_0 x^m + a_1 x^{m-1} + \cdots + a_m.$$

The forward difference of this function is

$$\Delta p_m(x) = p_m(x + h) - p_m(x)$$

and since

$$p_m(x + h) = a_0(x + h)^m + a_1(x + h)^{m-1} + \cdots + a_m$$

we have

$$\Delta p_m(x) = a_0\{(x + h)^m - x^m\} + \text{terms of lower degree}.$$

In particular the term in x^m disappears so that $\Delta p_m(x)$ is a polynomial of degree $m - 1$. In a similar manner the term in x^{m-1} will disappear when forming $\Delta(\Delta p_m(x))$, i.e. $\Delta^2 p_m(x)$ is a polynomial of degree $m - 2$. Repeated application of this process shows that $\Delta^m p_m(x)$ is a polynomial of degree 0, i.e. a constant. Thus the mth-order differences of an mth-degree polynomial are constant. The same result is obtained when using the backward difference operator ∇ and central difference operator δ. It is instructive for the reader to form a finite difference table for any cubic polynomial, e.g. form a difference table for $y = x^3 - 3x^2 + 2x + 1$ for $x = 0.0(0.2)1.0$. All of the entries in the third difference column will have the same value.

Suppose now that we form a finite difference table for a given set of data and find that the third-order differences, for example, are constant to within round-off error. The above result suggests that it is reasonable to represent the function with a cubic. More generally, if we find that the mth-order differences are constant to within round-off error we may represent the function with a polynomial of degree m.

3.2 INTERPOLATION FORMULAE INVOLVING FORWARD AND BACKWARD DIFFERENCES

From equation (3.29) we now present a simple but non-rigorous derivation of interpolation formulae involving forward and backward finite differences. A more rigorous derivation is described at the end of section 3.4. To express $f(x) = f(x_j + rh) = E^r f_j$ in terms of forward differences we substitute $E = 1 + \Delta$ from equation (3.20):

$$f(x) = f(x_j + rh) = E^r f_j = (1 + \Delta)^r f_j$$

$$= \left\{ 1 + r\Delta + \frac{r(r-1)}{2!} \Delta^2 + \cdots \right\} f_j,$$

i.e.

$$f(x) = f(x_j + rh)$$

$$= f_j + r\Delta f_j + \frac{r(r-1)}{2!} \Delta^2 f_j + \cdots + \frac{r(r-1)\cdots(r-n+1)}{n!} \Delta^n f_j + e_n(r), \quad (3.30)$$

where $e_n(r)$ is the error resulting from terminating the series after the nth-order difference $\Delta^n f_j$. In a similar manner we can express $f(x) = f(x_j + rh) = E^r f_j$ in terms of backward differences by substituting $E = (1 - \nabla)^{-1}$ from equation (3.21):

$$f(x) = f(x_j + rh) = E^r f_j = (1 - \nabla)^{-r} f_j$$

$$= \left\{ 1 + r\nabla + \frac{r(r-1)}{2!} \nabla^2 + \cdots \right\} f_j,$$

i.e.

$$f(x) = f(x_j + rh)$$

$$= f_j + r\nabla f_j + \frac{r(r+1)}{2!} \nabla^2 f_j + \cdots \frac{r(r+1)\cdots(r+n-1)}{n!} \nabla^n f_j + e_n(r). \quad (3.31)$$

Equations (3.30), (3.31) are called the Gregory–Newton forward and backward interpolation formulae respectively. The nth-degree polynomial, obtained by omitting $e_n(r)$ from the Gregory–Newton forward formula, passes through the points (x_k, f_k) for $k = j(1)j + n$. When $e_n(r)$ is omitted from the Gregory–Newton backward formula the resulting nth-degree polynomial passes through the points (x_k, f_k) for $k = j(-1)j - n$. These statements are not justified by the simple derivations of the formula presented here but are an immediate consequence of the derivation of the Gregory–Newton formulae from Newton's divided difference formula (section 3.4). However, the next example demonstrates these statements for the case $n = 2$.

Example 3.4 The functions

$$p_2(r) = f_4 + r\Delta f_4 + \frac{r(r-1)}{2!}\Delta^2 f_4$$

and

$$q_2(s) = f_6 + s\nabla f_6 + \frac{s(s+1)}{2!}\nabla^2 f_6,$$

where r and s are defined by $x = x_4 + rh = x_6 + sh$, give quadratic approximations to $f(x_4 + rh)$ and $f(x_6 + sh)$. Show that they represent the quadratic passing through

$$(x_4, f_4), (x_5, f_5), (x_6, f_6).$$

From the relationship $x = x_4 + rh = x_6 + sh$ we see that, when $x = x_4$, r takes the value zero and $s = (x_4 - x_6)/h = -2$. Substituting these values into the above formulae gives

$$p_2(0) = f_4$$
$$q_2(-2) = f_6 - 2\nabla f_6 + \nabla^2 f_6$$
$$= f_6 - 2(f_6 - f_5) + (f_6 - 2f_5 + f_4)$$
$$= f_4$$

showing that they both pass through (x_4, f_4). Similarly when $x = x_5$ we have

$$r = \frac{x_5 - x_4}{h} = 1, \quad s = \frac{x_5 - x_6}{h} = -1$$

and substituting these values into the formulae for $p_2(r)$ and $q_2(s)$ gives

$$p_2(1) = f_4 + \Delta f_4$$
$$= f_4 + (f_5 - f_4) = f_5$$
$$q_2(-1) = f_6 - \nabla f_6$$
$$= f_6 - (f_6 - f_5) = f_5.$$

Thus both quadratics pass through (x_5, f_5). Finally when $x = x_6$ we see that $s = 0$ and $r = 2$ and

$$p_2(2) = f_4 + 2\Delta f_4 + \Delta^2 f_4$$
$$= f_4 + 2(f_5 - f_4) + (f_6 - 2f_5 + f_4)$$
$$= f_6$$
$$q_2(0) = f_6,$$

i.e. both quadratics pass through (x_6, f_6).

More generally, there is a unique polynomial passing through a given set of points and the Gregory–Newton formulae give two alternative forms of this polynomial. Later we meet other forms of this unique polynomial.

To extend the above example a little, note that

$$p_3(r) = p_2(r) + \frac{r(r-1)(r-2)}{3!} \Delta^3 f_4$$

passes through the same points as $p_2(r)$, namely (x_4, f_4), (x_5, f_5), (x_6, f_6), and the next forward points of the set, (x_7, f_7). However

$$q_3(s) = q_2(s) + \frac{s(s+1)(s+2)}{3!} \nabla^3 f_6$$

passes through (x_4, f_4), (x_5, f_5), (x_6, f_6) and the next backward point of the set (x_3, f_3). In general, for each additional term included, the resulting polynomial passes through the next point of the set—forward or backward—according to the type of differences being used.

The pattern formed in a finite difference table by the finite differences from a given interpolation formula is called the associated formula pattern. The sets of differences in the Gregory–Newton forward and backward formulae are $\{f_j, \Delta f_j, \Delta^2 f_j, \ldots\}$ and $\{f_j, \nabla f_j, \nabla^2 f_j, \ldots\}$ respectively and the associated formula patterns are therefore Δ-bands and ∇-bands respectively (see Tables 3.2, 3.3). When estimating $f(x) = f(x_j + rh)$, having located x_j, the closest tabular point to x, the associated formula pattern should be marked on the numerical table. This simplifies the substitution of numbers from the table into the formula.

Because of their formula patterns the Gregory–Newton forward and backward interpolation formulae are most useful when interpolating near the top and bottom of a finite difference table respectively. More specifically, when x_j is near the bottom of a table, the corresponding Δ-band quickly runs out of the table, i.e. an insufficient number of f_j, Δf_j, $\Delta^2 f_j \ldots$ exist and consequently the Gregory–Newton forward formula cannot be used. Similarly the Gregory–Newton backward formula is unsuitable for use near the top of a difference table. It is emphasized that the Gregory–Newton formulae can only be used for interpolation when the tabular x-values are equally spaced. This restriction applies to all interpolation formulae involving finite differences because the formation of finite differences is based on the assumption that there is an equal spacing between the x-values. A more general interpolation formula for unequally spaced data is derived in section 3.4.

Example 3.5 A function $f(x)$ takes the values 0.47943, 0.64422, 0.78333, 0.89121, 0.96356, 0.99749, for $x = 0.5(0.2)1.5$. From these data, estimate $f(0.55)$.

The first step is to construct a finite difference table for the given data on page 43.

For a given set of data the finite difference table is unique. Forward, backward or central difference notation can be used to head the columns regardless of the type of differences to be used later for interpolation.

To estimate $f(0.55)$ we note that the closest tabular x-value to $x = 0.55$ is 0.5. Hence we set $x_0 = 0.5$ and since this is at the top of the table we use the Gregory–Newton forward formula:

$$f(x) = f(x_0 + rh) = f_0 + r\Delta f_0 + \frac{r(r-1)}{2!} \Delta^2 f_0 + \cdots.$$

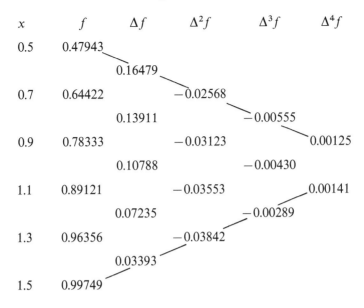

x	f	Δf	$\Delta^2 f$	$\Delta^3 f$	$\Delta^4 f$
0.5	0.47943				
		0.16479			
0.7	0.64422		−0.02568		
		0.13911		−0.00555	
0.9	0.78333		−0.03123		0.00125
		0.10788		−0.00430	
1.1	0.89121		−0.03553		0.00141
		0.07235		−0.00289	
1.3	0.96356		−0.03842		
		0.03393			
1.5	0.99749				

The differences needed for this formula lie on a Δ-band and are marked in the table to reduce the chances of error when substituting from the table into the formula. Since $x = x_0 + rh$ we have

$$0.55 = 0.5 + r(0.2),$$

giving $r = 0.25$. Substitution into the Gregory–Newton formula gives

$$f(0.55) = 0.47493 + 0.04120 + 0.00241 - 0.00030 - 0.00005 +$$

and we see that $f(0.55) = 0.5227$ to four decimal places.

To estimate $f(1.47)$ we note that the closest tabular x-value is $x_5 = 1.5$ and since this is at the bottom of the table we must use the Gregory–Newton backward formula:

$$f(x) = f(x_5 + rh) = f_5 + r\nabla f_5 + \frac{r(r+1)}{2!}\nabla^2 f_5 + \cdots.$$

The differences needed for this formula lie on the marked ∇-band. Since $x = x_5 + rh$ we have

$$1.47 = 1.5 + r(0.2),$$

giving $r = -0.15$ and the reader is left to check that substitution into the Gregory–Newton backward formula leads to $f(1.47) = 0.9949$ to four decimal places.

Algorithm 3.1. The Gregory–Newton forward formula to interpolate using a table with uniformly spaced points

The initial tabular point is $x0$ and the n tabular function values are held in $y[0], \ldots, y[n-1]$, with h the step length. The degree of the approximating polynomial is m and the required difference is held in Diff[0]. The interpolated value of y at the point x is yx.

```
read(x0, h, n);
FOR i:= 0 TO n − 1 DO read(y[i]);
read(x, m);
j:= round((x − x0)/h − 0.5);
r:= (x − x0)/h − j;
IF (j > = 0) AND (j < n − m) THEN
   BEGIN
   FOR i:= 0 TO m DO Diff[i]:= y[j + i + 1] − y[j + i];
   Coeff:= r;
   yx:= y[j] + Coeff*Diff[0];
   FOR i:= 2 TO m DO
      BEGIN
      Coeff:= Coeff*(r − i + 1)/i;
      FOR j:= 0 TO m − i DO Diff[j]:= Diff[j + 1] − Diff[j];
      yx:= yx + Coeff*Diff[0]
      END;
   write (yx)
   END
ELSE write ('not enough function values').
```

Algorithm 3.2. The Gregory–Newton backward formula to interpolate using a table with uniformly spaced points

The initial tabular point is $x0$ and the n tabular function values are held in $y[0], \ldots, y[n − 1]$, with h the step length. The degree of the approximating polynomial is m and the required difference is held in Diff[0]. The interpolated value of y at the point x is yx.

```
read(x0, h, n);
FOR i:= 0 TO n − 1 DO read(y[i]);
read(x, m);
j:= round((x − x0)/h + 0.5);
r:= (x − x0)/h − j;
IF (j > = m) AND (j < n) THEN
   BEGIN
   FOR i:= 0 DOWNTO − m + 1 DO Diff[i]:= y[j + i] − y[j + i − 1];
   Coeff:= r;
   yx:= y[j] + Coeff*Diff[0];
   FOR i:= 2 TO m DO
      BEGIN
      Coeff:= Coeff*(r + i − 1)/i;
      FOR j:= 0 DOWNTO − m + i DO Diff[j]:= Diff[j] − Diff[j − 1];
      yx:= yx + Coeff*Diff[0]
      END;
   write(yx)
   END
ELSE write('not enough function values').
```

The Gregory–Newton formulae incorporate data that always lie to one side of the point of interpolation. This is advantageous when interpolating near the top and bottom of a finite difference table as demonstrated in the above example. Although both formulae can normally be used in the central area of a finite difference table it is preferable to use a formula that incorporates data on each side of the point of interpolation. Therefore, our next aim is to construct interpolation formulae with associated difference patterns that are, as far as possible, symmetrical with respect to the point of interpolation. These formulae involve central differences. The complicated relationship between E and δ (equation (3.22)) does not permit a simple derivation of central difference formulae. We will derive the central difference formulae from a more general interpolation formula which is presented in the following section.

3.3 DIVIDED DIFFERENCES AND NEWTON'S DIVIDED DIFFERENCE FORMULA

We only present a limited discussion of divided differences and the practical use of Newton's divided difference formula. The main reason for its inclusion is to derive formulae involving finite differences, central differences in particular (see section 3.4). However, it does have a practical use in its own right. The formation of finite differences in a table is based on the assumption that the tabular x-values are equally spaced, since finite differences are only defined for equally spaced data. Consequently interpolation formulae involving finite differences can only be used when the x-data are equally spaced. However, Newton's divided difference formula can be used with any data regardless of the spacing of the x-values.

Divided differences

Table 3.5 illustrates the square-bracket notation, used for divided differences and, for consistency, we write $f[x_i]$ instead of $f(x_i)$ when using divided differences. Note that the order of a divided difference is always one less than the number of x-values within the square brackets.

The formation of a divided difference table from given data is a simple extension of the formation of a finite difference table. Table 3.5 shows how each column is calculated from left to right in terms of the previous column. Insertion of the correct x-values into the denominators can be achieved by drawing (by eye in practice) diagonals. These have been indicated in the table for $f[x_{-1}, x_0, x_1, x_2]$. Not only can the x-values have unequal spacing when forming divided differences but they can also be in any numerical order. Thus, although we normally order the data such that the x-values are increasing, this is not essential when forming a divided difference table.

Now that we have introduced divided differences and indicated how they are formed in a table we give the formal recursive definition of divided differences:

$$f[x_j] = f(x_j), \quad f[x_j, x_{j+1}] = \frac{f[x_{j+1}] - f[x_j]}{x_{j+1} - x_j}$$

$$f[x_j, x_{j+1}, \ldots, x_{j+k}] = \frac{f[x_{j+1}, \ldots, x_{j+k}] - f[x_j, \ldots, x_{j+k-1}]}{x_{j+k} - x_j}. \tag{3.32}$$

Table 3.5 Divided difference notation

x	$f(x)$	First divided differences	Second divided differences	Third divided differences
x_{-2}	$f[x_{-2}]$			
		$f[x_{-2},x_{-1}]=\dfrac{f[x_{-1}]-f[x_{-2}]}{x_{-1}-x_{-2}}$		
x_{-1}	$f[x_{-1}]$		$f[x_{-2},x_{-1},x_0]=\dfrac{f[x_{-1},x_0]-f[x_{-2},x_{-1}]}{x_0-x_{-2}}$	
		$f[x_{-1},x_0]=\dfrac{f[x_0]-f[x_{-1}]}{x_0-x_{-1}}$		$f[x_{-2},x_{-1},x_0,x_1]=\dfrac{f[x_{-1},x_0,x_1]-f[x_{-2},x_{-1},x_0]}{x_1-x_{-2}}$
x_0	$f[x_0]$		$f[x_{-1},x_0,x_1]=\dfrac{f[x_0,x_1]-f[x_{-1},x_0]}{x_1-x_{-1}}$	
		$f[x_0,x_1]=\dfrac{f[x_1]-f[x_0]}{x_1-x_0}$		$f[x_{-1},x_0,x_1,x_2]=\dfrac{f[x_0,x_1,x_2]-f[x_{-1},x_0,x_1]}{x_2-x_{-1}}$
x_1	$f[x_1]$		$f[x_0,x_1,x_2]=\dfrac{f[x_1,x_2]-f[x_0,x_1]}{x_2-x_0}$	
		$f[x_1,x_2]=\dfrac{f[x_2]-f[x]}{x_2-x_1}$		$f[x_0,x_1,x_2,x_3]=\dfrac{f[x_1,x_2,x_3]-f[x_0,x_1,x_2]}{x_3-x_0}$
x_2	$f[x_2]$		$f[x_1,x_2,x_3]=\dfrac{f[x_2,x_3]-f[x_1,x_2]}{x_3-x_1}$	
		$f[x_2,x_3]=\dfrac{f[x_3]-f[x_2]}{x_3-x_2}$		
x_3	$f[x_3]$			

In particular we will make use of the result that the value of a divided difference does not depend on the order in which the x-values appear in the divided difference. We verify this for $f[x_j, x_{j+1}, x_{j+2}]$. From the definition (3.32) we have

$$f[x_j, x_{j+1}, x_{j+2}] = \frac{f[x_{j+1}, x_{j+2}] - f[x_j, x_{j+1}]}{x_{j+2} - x_j}$$

and substituting for $f[x_{j+1}, x_{j+2}]$ and $f[x_j, x_{j+1}]$ leads to

$$f[x_j, x_{j+1}, x_{j+2}] = \frac{f(x_j)}{(x_j - x_{j+1})(x_j - x_{j+2})} + \frac{f(x_{j+1})}{(x_{j+1} - x_j)(x_{j+1} - x_{j+2})}$$

$$+ \frac{f(x_{j+2})}{(x_{j+2} - x_j)(x_{j+2} - x_{j+1})}. \tag{3.33}$$

Interchanging x_j and x_{j+1}, for example, shows that the right-hand side of (3.33) does not depend on the order of x_j, x_{j+1}, x_{j+2}. Consequently the value of $f[x_j, x_{j+1}, x_{j+2}]$ does not depend on the order in which x_j, x_{j+1}, x_{j+2} appear. More generally it can be shown that

$$f[x_j, x_{j+1}, \ldots, x_{j+k}] = \sum_{i=j}^{j+k} \frac{f(x_i)}{\prod_i}, \tag{3.34}$$

where

$$\prod_i = (x_i - x_j)(x_i - x_{j+1}) \cdots (x_i - x_{i-1})(x_i - x_{i+1}) \cdots (x_i - x_{j+k}).$$

Notice that $\prod_i$ is the product of all factors of the form $(x_i - x_r)$ for $r = j, j+1, \ldots, j+k$. but excluding $r = i$. Equation (3.34) shows that the value of $f[x_j, x_{j+1}, \ldots, x_{j+k}]$ does not depend on the order of $x_j, x_{j+1}, \ldots, x_{j+k}$. The proof of (3.34) involves cumbersome algebra and mathematical induction with respect to k (Exercise 16).

Newton's divided difference polynomial

This is a special form of the polynomial passing though a given set of data points. To derive it we determine the values of the $n+1$ constants $a_0, a_1, \ldots, a_n$ such that the nth-degree polynomial.

$$p_n(x) = a_0 + (x - x_0)a_1 + (x - x_0)(x - x_1)a_2 + (x - x_0)(x - x_1)(x - x_2)a_3$$
$$+ \cdots + (x - x_0)(x - x_1) \cdots (x - x_{n-1})a_n \tag{3.35}$$

passes through the $n+1$ points (x_j, f_j) for $j = 0(1)n$. Thus we require

$$p_n(x_j) = f(x_j), \quad j = 0(1)n. \tag{3.36}$$

Since (3.35) will be used to approximate $f(x)$ we write

$$f(x) = a_0 + (x - x_0)a_1 + (x - x_0)(x - x_1)a_2$$
$$+ \cdots + (x - x_0)(x - x_1) \cdots (x - x_{n-1})a_n + e_n(x), \tag{3.37}$$

i.e. $e_n(x) = f(x) - p_n(x)$ represents the error in the approximation of $f(x)$ by $p_n(x)$ when $x \neq x_j$. Obviously the error is zero when $x = x_j$, i.e. the coefficients $a_0, a_1, a_2, \ldots$ are chosen such that (3.36) is satisfied in which case

$$e_n(x_j) = 0, \quad j = 0(1)n. \tag{3.38}$$

We determine the coefficients $a_0, a_1, a_2, \ldots$ by setting $x = x_0, x_1, x_2, \ldots$ in turn into (3.37), using (3.38). Thus setting $x = x_0$ in (3.37) we see immediately that $a_0 = f(x_0)$. To determine a_1 we move a_0 to the left-hand side of (3.37), substitute $a_0 = f(x_0)$ and divide through by $(x - x_0)$:

$$\frac{f(x) - f(x_0)}{x - x_0} = a_1 + (x - x_1)a_2 + (x - x_1)(x - x_2)a_3$$

$$+ \cdots + (x - x_1) \cdots (x - x_{n-1})a_n + \frac{e_n(x)}{(x - x_0)}. \tag{3.39}$$

The left-hand side of (3.39) is $f[x, x_0]$ and setting $x = x_1$ gives $a_1 = f[x_1, x_0] = f[x_0, x_1]$. We now rearrange (3.39) into the form

$$\frac{f[x, x_0] - f[x_0, x_1]}{x - x_1} = a_2 + (x - x_2)a_3 + \cdots + (x - x_2) \cdots (x - x_{n-1})a_n$$

$$+ \frac{e_n(x)}{(x - x_0)(x - x_1)}. \tag{3.40}$$

The right-hand side is $f[x, x_0, x_1]$ and setting $x = x_2$ gives $a_2 = f[x_2, x_0, x_1] = f[x_0, x_1, x_2]$. Proceeding in this manner we find that in general

$$a_k = f[x_0, x_1, \ldots, x_k], \quad k = 1(1)n, \tag{3.41}$$

and substituting back into (3.37) gives

$$f(x) = f[x_0] + (x - x_0)f[x_0, x_1] + (x - x_0)(x - x_1)f[x_0, x_1, x_2]$$
$$+ (x - x_0)(x - x_1)(x - x_2)f[x_0, x_1, x_2, x_3]$$
$$+ (x - x_0)(x - x_1) \cdots (x - x_{n-1})f[x_0, x_1, x_2, \ldots, x_n] + e_n(x). \tag{3.42}$$

This is Newton's divided difference formula. Omission of $e_n(x)$ leaves Newton's divided difference polynomial, $p_n(x)$, which passes through the $n + 1$ points (x_k, f_k) for $k = 0(1)n$.

If a, b, c are any three numbers, the quadratic passing through $(a, f(a))$, $(b, f(b))$ and $(c, f(c))$ can be expressed in the form

$$p_2(x) = f[a] + (x - a)f[a, b] + (x - a)(x - b)f[a, b, c].$$

This emphasizes that the order of the x-data is irrelevant in Newton's divided difference polynomial. Also by adding in one extra term, the polynomial passes through one extra point, e.g.

$$p_3(x) = p_2(x) + (x - a)(x - b)(x - c)f[a, b, c, d]$$

passes through $(d, f(d))$ as well as the three points that $p_2(x)$ passes through.

Example 3.6 For $x = 0.1, 0.3, 0.4, 0.7, 0.9$, a function $f(x)$ takes the values 1.10517, 1.34989, 1.49187, 2.01390, 2.45985. Form a table of divided differences from these data and estimate a value for $f(0.5)$.

The table of divided differences is

x f
0.1   1.10517

$f[0.1,0.3]$
1.22360

$f[0.1,0.3,0.4]$

0.3   1.34989

0.65400

$f[0.3,0.4]$ $f[0.1,0.3,0.4,0.1]$
1.41980   0.24458

$f[0.3,0.4,0.7]$

0.4   1.49187 $f[0.1,0.3,0.4,0.7,0.9]$

0.80075   0.06625

$f[0.4,0.7]$ $f[0.3,0.4,0.7,0.9]$
1.74010   0.29758

$f[0.4,0.7,0.9]$

0.7   2.0139

0.97930

$f[0.7,0.9]$
2.22975

0.9   2.45985

It is not normally necessary to include the notation for the differences in a divided difference table, but here it is included to help the reader check the values of the differences. To estimate $f(0.5)$ from this table we make use of the fact that the x-values can be introduced into Newton's divided difference polynomial in any order by first introducing those closest to $x = 0.5$. Thus we write Newton's divided difference polynomial in the form

$$f(x) = f[0.4] + (x - 0.4)f[0.4,0.7] + (x - 0.4)(x - 0.7)f[0.4,0.7,0.3]$$
$$+ (x - 0.4)(x - 0.7)(x - 0.3)f[0.4,0.7,0.3,0.9]$$
$$+ (x - 0.4)(x - 0.7)(x - 0.3)(x - 0.9)f[0.4,0.7,0.3,0.9,0.1].$$

Note that this does represent the polynomial of degree four passing though the five given data points. Setting $x = 0.5$ gives

$$f(0.5) = 1.49187 + 0.17401 - 0.016202 - 0.00119 + 0.00011 = 1.64878.$$

Since the last term affects the fourth decimal place we can only rely on the answer to three decimal places, i.e. $f(0.5) = 1.649$. However, it is possible that the answer is accurate to four decimal places, namely $f(0.5) = 1.6489$, since the last term, 0.00011, only makes a small contribution to the fourth decimal place and successive terms are decreasing by a factor of ten approximately.

Relation between finite differences and divided differences

For equally spaced x-data ($x_{k+1} = x_k + h$, where h is the constant spacing) we can relate finite differences and divided differences. We show that

$$f[x_j, x_{j+1}, \ldots, x_{j+k-1}, x_{j+k}] = \frac{1}{k!} \frac{1}{h^k} \Delta^k f_j, \tag{3.43}$$

$$= \frac{1}{k!} \frac{1}{h^k} \nabla^k f_{j+k}, \tag{3.44}$$

$$= \frac{1}{k!} \frac{1}{h^k} \delta^k f_{j+k/2}. \tag{3.45}$$

We prove the first by induction with respect to the number of x-values in the divided difference. By definition

$$f[x_j, x_{j+1}] = \frac{f(x_{j+1}) - f(x_j)}{x_{j+1} - x_j} = \frac{1}{h} \Delta f_j,$$

showing that (3.43) holds for $k = 1$. Assuming that (3.43) is true for any value of j and with $k = n$ we have

$$f[x_j, x_{j+1}, \ldots, x_{j+n}] = \frac{1}{n!} \frac{1}{h^n} \Delta^n f_j$$

and, replacing j by $j + 1$,

$$f[x_{j+1}, x_{j+2}, \ldots, x_{j+n+1}] = \frac{1}{n!} \frac{1}{h^n} \Delta^n f_{j+1}.$$

We now show that (3.43) is also true when $k = n + 1$. Setting $k = n + 1$ in the left-hand side of (3.43), using the definition of divided differences and substituting from the last two formulae lead to

$$f[x_j, x_{j+1}, \ldots, x_{j+n}, x_{j+n+1}]$$

$$= \frac{f[x_{j+1}, \ldots, x_{j+n+1}] - f[x_j, \ldots, x_{j+n}]}{x_{j+n+1} - x_j}$$

$$= \frac{1}{(n+1)h} \left[\frac{1}{n!} \frac{1}{h^n} \Delta^n f_{j+1} - \frac{1}{n!} \frac{1}{h^n} \Delta^n f_j \right]$$

$$= \frac{1}{(n+1)! h^{n+1}} \Delta^n (f_{j+1} - f_j)$$

$$= \frac{1}{(n+1)! h^{n+1}} \Delta^{n+1} f_j.$$

The last expression is the right-hand side of (3.43) with $k = n + 1$. Consequently (3.43) must hold for all values of k. Equations (3.44) and (3.45) follow almost immediately from (3.43) and the operator relations

$$\Delta = \delta E^{1/2} = \nabla E,$$

which are easily derived from the definitions of E, Δ and δ.

Alternative derivation of forward and backward difference interpolation formulae

From Table 3.5 it can be seen that the divided differences which appear in Newton's divided difference formula (3.42) lie on a Δ-band. In particular the formula pattern associated with (3.42) is identical to the formula pattern associated with the Gregory–Newton forward formulae (3.30). Further, Newton's divided difference formula passing through the points (x_k, f_k) for $k = j(1)j + n$ is (compare with (3.42))

$$f(x) = f[x_j] + (x - x_j)f[x_j, x_{j+1}] + (x - x_j)(x - x_{j+1})f[x_j, x_{j+1}, x_{j+2}]$$
$$+ \cdots + (x - x_j)(x - x_{j+1}) \cdots (x - x_{j+n-1})f[x_j, x_{j+1}, \ldots, x_{j+n}] + e_n(x). \quad (3.46)$$

For equally spaced x-data this is equivalent to (3.30) and leads to the derivation of (3.30). The divided differences in (3.42) can be replaced by forward differences using (3.43) with $k = 1, 2, 3, \ldots$ in turn. The factors such as $(x - x_j)$ and $(x - x_{j+1})$ can be replaced using the relationships $x = x_j + rh$, $x_{j+1} = x_j + h$, $x_{j+2} = x_j + 2h$, etc. It is left for the reader to check that this leads directly to equation (3.30). An immediate consequence of this derivation is that the nth-degree Gregory–Newton forward polynomial, obtained by omitting $e_n(r)$ from (3.30), passes through the points (x_k, f_k) for $k = j(1)j + n$.

To derive the Gregory–Newton backward formula (3.31) from Newton's divided difference formula we first observe the divided differences that make up the same formula pattern as the backward differences in (3.31). These backward differences lie on a ∇-band starting at f_j in a finite difference table. From the following section of a divided difference table we see that the divided differences on the equivalent band are $f[x_j], f[x_{j-1}, x_j], \ldots, f[x_{j-n}, x_{j-n+1}, \ldots, x_{j-1}, x_j]$.

$$
\begin{array}{llll}
x_{j-3} & f[x_{j-3}] & & \\
 & & f[x_{j-3}, x_{j-2}] & \\
x_{j-2} & f[x_{j-2}] & & f[x_{j-3}, x_{j-2}, x_{j-1}] \\
 & & f[x_{j-2}, x_{j-1}] & & f[x_{j-3}, x_{j-2}, x_{j-1}, x_j] \\
x_{j-1} & f[x_{j-1}] & & f[x_{j-2}, x_{j-1}, x_j] \\
 & & f[x_{j-1}, x_j] & \\
x_j & f[x_j] & &
\end{array}
$$

Consequently we introduce the x-data into Newton's divided difference formula in the order $x_j, x_{j-1}, \ldots, x_{j-n}$ leading to

$$f(x) = f[x_j] + (x - x_j)f[x_j, x_{j-1}] + (x - x_j)(x - x_{j-1})f[x_j, x_{j-1}, x_{j-2}]$$
$$+ \cdots + (x - x_j)(x - x_{j-1}) \cdots (x - x_{j-n+1})f[x_j, x_{j-1}, \ldots, x_{j-n}] + e_n(x). \quad (3.47)$$

Note that we have made use of the fact that the value of divided difference does not depend on the order of the x-data. Equation (3.31) can be derived from (3.47) by using (3.44) and $x = x_j + rh$. Again the details are left to the reader. However, a slightly different form of (3.44) is required. Noting that j can take any value in (3.44) we replace it by $j - k$ and reverse the order of the x-data to obtain

$$f(x_j, x_{j-1}, \ldots, x_{j-k}) = \frac{1}{k!} \frac{1}{h^k} \nabla^k f_j. \quad (3.48)$$

3.4 CENTRAL DIFFERENCE INTERPOLATION FORMULAE

By considering the formula patterns associated with the forward and backward
difference interpolation formulae, (3.30) and (3.31), we see that the data involved in
these formulae lie to one side of the point of interpolation. It is preferable to use
formulae that involve differences that are close to the point of interpolation and
symmetrically spaced about the point of interpolation. Consequently we see a need to
construct interpolation formulae with formula patterns 3 and 4 shown in Table 3.6.
Pattern 3 is suitable when the point of interpolation is closer to a tabular point than a
mid-point since the data involved are located almost symmetrically about the point of
interpolation. When the point of interpolation is closer to a mid-point than a tabular
point, pattern 4 is most suitable, again because the data involved are located almost
symmetrically about the point of interpolation.

We derive the central difference formulae with the patterns shown in Table 3.6 from
suitable forms of Newton's divided difference formula. Because Newton's divided
difference formula only contains one difference of each order we must first derive the
formulae whose associated formula patterns are patterns 1 and 2 in Table 3.6. These are

Table 3.6 Formula patterns for central difference interpolation
formula.

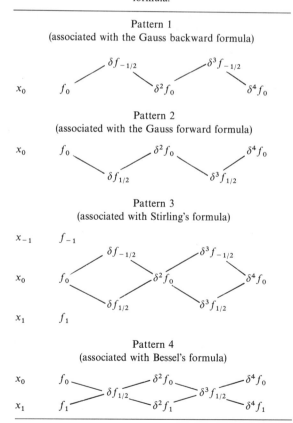

only derived as an intermediate step towards obtaining Stirling's formula and Bessel's formula. To obtain the appropriate forms of Newton's divided difference formula we look at the divided differences associated with patterns 1 and 2 in a divided difference table. The required section of the divided difference table is (see Table 3.5)

$$
\begin{array}{llll}
x_{-1} & f[x_{-1}] \\
 & & f[x_{-1}, x_0] & & f[x_{-2}, x_{-1}, x_0, x_1] \\
x_0 & f[x_0] & & f[x_{-1}, x_0, x_1] \\
 & & f[x_0, x_1] & & f[x_{-1}, x_0, x_1, x_2] \\
x_1 & f[x_1]
\end{array}
$$

For pattern 1, following the x-values through the divided differences, we must introduce the x-data into Newton's divided differences formula in the order $x_0, x_{-1}, x_1, x_{-2}, x_2, x_{-3}, x_3, \ldots$. Similarly, for pattern 2, the x-data must be introduced into Newton's divided difference formula in the order $x_0, x_1, x_{-1}, x_2, x_{-2}, x_3, x_{-3}, \ldots$. These sequences have been extended to emphasize the alternating indices; only the first four values of each sequence are evident from the table above. Thus for pattern 2, Newton's divided difference formula takes the form

$$
\begin{aligned}
f(x) = {}& f[x_0] + (x - x_0)f[x_0, x_1] + (x - x_0)(x - x_1)f[x_{-1}, x_0, x_1] \\
& + (x - x_0)(x - x_1)(x - x_{-1})f[x_{-1}, x_0, x_1, x_2] \\
& + (x - x_0)(x - x_1)(x - x_{-1})(x - x_2)f[x_{-2}, x_{-1}, x_0, x_1, x_2] + \cdots. \quad (3.49)
\end{aligned}
$$

We use (3.45) to express the divided differences in terms of central differences. The even-order differences have been expressed in the form $f[x_{-p}, \ldots, x_0, \ldots, x_p]$. Equating the smallest and largest suffix to the corresponding suffices in (3.45) gives

$$
j = -p, \quad j + k = p.
$$

Thus $j = -p$, $k = 2p$ and from (3.45)

$$
f[x_{-p}, \ldots, x_0, \ldots, x_p] = \frac{1}{(2p)!} \frac{1}{h^{2p}} \delta^{2p} f_0. \quad (3.50)
$$

In a similar manner it is easily shown that the odd-order differences in (3.49) can be expressed in the form

$$
f[x_{-p}, \ldots, x_0, \ldots, x_p, x_{p+1}] = \frac{1}{(2p + 1)!} \frac{1}{h^{2p+1}} \delta^{2p+1} f_{1/2}. \quad (3.51)
$$

In addition, since we are using finite differences, the x-values are ordered and equally spaced and we may write

$$
x = x_0 + rh
$$
$$
x_{\pm 1} = x_0 \pm h, \quad x_{\pm 2} = x_0 \pm 2h, \cdots,
$$

leading to

$$
x - x_0 = rh, \quad x - x_1 = (r - 1)h, \quad (x - x_2) = (r - 2)h, \cdots
$$
$$
x - x_{-1} = (r + 1)h, \quad (x - x_{-2}) = (r + 2)h, \cdots. \quad (3.52)
$$

Using (3.50), (3.51) with $p = 0, 1, 2, \ldots$ in turn and (3.52), equation (3.49) can be expressed in the form

$$f(x) = f(x_0 + rh)$$

$$= f_0 + \frac{r}{1!}\delta f_{1/2} + \frac{r(r-1)}{2!}\delta^2 f_0 + \frac{r(r-1)(r+1)}{3!}\delta^3 f_{1/2}$$

$$+ \frac{r(r-1)(r+1)(r-2)}{4!}\delta^4 f_0 \cdots. \qquad (3.53)$$

This is sometimes called the Gauss forward interpolation formula, and the associated formula pattern is pattern 2 in Table 3.6.

As indicated above, to obtain a similar formula associated with pattern 1, the starting point in Newton's divided difference formula with x-data introduced in the order $x_0, x_{-1}, x_1, x_2, x_{-2}, \ldots$. This leads to the Gauss backward formula

$$f(x) = f(x_0 + rh)$$

$$= f_0 + \frac{r}{1!}\delta f_{-1/2} + \frac{r(r+1)}{2!}\delta^2 f_0 + \frac{r(r+1)(r-1)}{3!}\delta^3 f_{-1/2}$$

$$+ \frac{r(r+1)(r-1)(r+2)}{4!}\delta^4 f_0 + \cdots. \qquad (3.54)$$

The detailed derivation is very similar to the derivation of (3.53).

Since both formulae are derived from alternative forms of Newton's divided difference formula, when terminated with a difference of order $2m$ both formulae pass through the data points (x_k, f_k) for $k = 0, \pm 1, \pm 2, \ldots, \pm m$. When terminated with a difference of order $2m + 1$ the Gauss forward and Gauss backward formulae pass through the additional points (x_{m+1}, f_{m+1}) and (x_{-m-1}, f_{-m-1}) respectively.

More general forms of (3.53) and (3.54) can be obtained as follows. If x_j is the closest tabular point to x then r is defined by $x = x_j + rh$ and

$$f(x) = f(x_j + rh) = E^j f(x_0 + rh).$$

Replacing $f(x_0 + rh)$ by the right-hand side (3.53) and observing that

$$E^j f_0 = f_j, \quad E^j \delta f_{1/2} = \delta f_{j+1/2}, \quad E^j \delta^2 f_0 = \delta^2 f_j, \text{ etc.}$$

leads to

$$f(x) = f(x_j + rh)$$

$$= f_j + \frac{r}{1!}\delta f_{j+1/2} + \frac{r(r-1)}{2!}\delta^2 f_j + \frac{2r(r-1)(r+1)}{3!}\delta^3 f_{j+1/2}$$

$$+ \cdots + \frac{r(r-1)(r+1)(r-2)}{4!}\delta^4 f_j + \cdots. \qquad (3.55)$$

Similarly if $f(x_0 + rh)$ is replaced by the right-hand side of (3.54) we obtain

$$f(x) = f(x_j + rh)$$

$$= f_j + \frac{r}{1!}\delta f_{j-1/2} + \frac{r(r+1)}{2!}\delta^2 f_j + \frac{r(r+1)(r-1)}{3!}\delta^3 f_{j-1/2}$$

$$+ \cdots + \frac{r(r+1)(r-1)(r+2)}{4!}\delta^4 f_j + \cdots. \qquad (3.56)$$

The associated formula patterns are patterns 1 and 2 but with j added to each suffix.

Stirling's formula

To obtain the central difference interpolation formula whose associated formula pattern is pattern 3 we simply form the mean of (3.53) and (3.54):

$$f(x) = f(x_0 + rh)$$

$$= f_0 + \frac{r}{1!}\left\{\frac{\delta f_{1/2} + \delta f_{-1/2}}{2}\right\} + \frac{r}{2!}\left\{\frac{(r-1)+(r+1)}{2}\right\}\delta^2 f_0$$

$$+ \frac{r(r-1)(r+1)}{3!}\left\{\frac{\delta^3 f_{1/2} + \delta^3 f_{-1/2}}{2}\right\}$$

$$+ \frac{r(r-1)(r+1)}{4!}\left\{\frac{(r-2)+(r+2)}{2}\right\}\delta^4 f_0 + \cdots. \tag{3.57a}$$

Using the averaging operator μ we may write

$$\mu\delta f_0 = \tfrac{1}{2}(\delta f_{1/2} + \delta f_{-1/2}), \quad \mu\delta^3 f_0 = \tfrac{1}{2}(\delta^3 f_{1/2} + \delta^3 f_{-1/2}), \ldots$$

and (3.57a) can be expressed in the form

$$f(x) = f(x_0 + rh)$$

$$= f_0 + \frac{r}{1!}\mu\delta f_0 + \frac{r^2}{2!}\delta^2 f_0 + \frac{r(r^2 - 1^2)}{3!}\mu\delta^3 f_0$$

$$+ \frac{r^2(r^2 - 1^2)}{4!}\delta^4 f_0 + \cdots. \tag{3.57b}$$

This is Stirling's interpolation formula. When x lies close to x_j, we define r by $x = x_j + rh$. Then, since $f(x) = f(x_j + rh) = E^j f(x_0 + rh)$, replacing $f(x_0 + rh)$ by the righthand side of (3.57) gives the required form of Stirling's formula which is (3.57) with j added to all of the suffices.

Since Stirling's formula is the average of the two Gauss formulae it will only pass through those points that both of the Gauss formulae pass through. We have already

Table 3.7 Range of values of k for (x_k, f_k) indicating the data points that each interpolation polynomial passes through

Order of highest difference retained	$2m$	$2m+1$
Gauss forward formula (3.53)	$k = -m(1)m$	$k = -m(1)m+1$
Gauss backward formula (3.54)	$k = -m(1)m$	$k = -m-1(1)m$
Stirling's formula (3.57)	$k = -m(1)m$	$k = -m(1)m$
Gauss backward formula (3.60), based at x_1	$k = -m+1(1)m+1$	$k = -m(1)m+1$
Bessel's formula (3.61)	$k = -m+1(1)m$	$k = -m(1)m+1$

noted, when each of the Gauss formulae (3.53), (3.54) are terminated with a difference order $2m$, they both pass through the set of points (x_k, f_k) for $k = -m(1)m$. Consequently Stirling's formula will also pass through (x_k, f_k) for $k = -m(1)m$ when terminated with a difference of order $2m$. However, when terminated with a difference of order $2m + 1$ the Gauss forward and backward formulae pass through the additional points (x_{m+1}, f_{m+1}) and (x_{-m-1}, f_{-m-1}) respectively as well as (x_k, f_k) for $k = -m(1)m$. Thus, when Stirling's formula is terminated with a difference of order $2m + 1$ it will only pass through the common points (x_k, f_k) for $k = -m(1)m$. This information is summarized in Table 3.7.

Bessel's formula

The formula patterns for (3.55) and (3.56) are

$$x_j \quad f_j \qquad\qquad \delta^2 f_j \qquad\qquad\qquad \delta^4 f_j$$
$$\delta f_{j+1/2} \qquad\qquad \delta^3 f_{j+1/2}$$

and

$$\delta f_{j-1/2} \qquad\qquad \delta^3 f_{j-1/2}$$
$$x_j \quad f_j \qquad \delta^2 f_j \qquad\qquad\qquad \delta^4 f_j$$

respectively. Taking $j = 0$ in the first pattern and $j = 1$ in the second pattern and superimposing the patterns leads to pattern 4 in Table 3.6. Therefore, to derive the corresponding interpolation formula we form the mean of equation (3.55) with $j = 0$ and (3.56) with $j = 1$. Equation (3.55) with $j = 0$ is already given in (3.53). Equation (3.56) with $j = 1$ is

$$f(x) = f(x_1 + sh)$$

$$= f_1 + \frac{s}{1!} \delta f_{1/2} + \frac{s(s+1)}{2!} \delta^2 f_1 + \frac{s(s+1)(s-1)}{3!} \delta^3 f_{1/2}$$

$$+ \frac{s(s+1)(s-1)(s+2)}{4!} \delta^4 f_1 + \cdots. \tag{3.58}$$

Note that r and s in (3.53) and (3.58) are distinct. They are defined by the linear relations

$$x = x_0 + rh = x_1 + sh, \tag{3.59}$$

i.e. r measures in units of h from x_0 whereas s measures in units of h from x_1. Putting $x_1 = x_0 + h$ in (3.59) we see that $s = r - 1$ and making this substitution in (3.58) gives

$$f(x) = f(x_0 + rh)$$

$$= f_1 + \frac{(r-1)}{1!} \delta f_{1/2} + \frac{(r-1)}{2!} \delta^2 f_1 + \frac{(r-1)r(r-2)}{3!} \delta^3 f_{1/2}$$

$$+ \frac{(r-1)r(r-2)(r+1)}{4!} \delta^4 f_1 + \cdots. \tag{3.60}$$

Finally, forming the mean of (3.53) and (3.60) leads to Bessel's interpolation formula,

$$f(x) = f(x_0 + rh)$$

$$= \mu f_{1/2} + (r - \tfrac{1}{2})\delta f_{1/2} + \frac{r(r-1)}{2!}\mu\delta^2 f_{1/2}$$

$$+ \frac{r(r-1)(r-1/2)}{3!}\delta^3 f_{1/2} + \frac{r(r-1)(r+1)(r-2)}{4!}\mu\delta^4 f_{1/2} + \cdots \qquad (3.61)$$

and the associated formula pattern is pattern 4 in Table 3.6. More generally, if x lies between x_j and x_{j+1}, we define r by $x = x_j + rh$. Then, since $f(x) = f(x_j + rh) = E^j f(x_0 + rh)$, operation of E^j on the right-hand side of (3.61) gives the required form of Bessel's formula which is (3.61) with j added to all of the suffices.

By considering the points that the Gauss formulae, (3.53) and (3.60), pass through we can determine the points that Bessel's formula passes through (see Table 3.7). When terminated after differences of order $2m + 1$, Bessel's formula passes through the $2m + 2$ points (x_k, f_k) for $k = -m(1)m + 1$. However, when terminated after differences of order $2m$, the resulting polynomial of degree $2m$ only passes through the $2m$ points (x_k, f_k) for $k = -m + 1(1)m$.

Example 3.7 Obtain estimates of $f(0.98)$ and $f(1.08)$ to four decimal places for the function specified numerically in Example 3.5.

The value $x = 0.98$ lies close to a mid-point and we therefore use Bessel's formula with $x_0 = 0.9$ to estimate $f(0.98)$ because the associated formula pattern is symmetrical about the mid-point 1.0. Noting Bessel's formula pattern in Table 3.6 the required section of the numerical table in Example 3.5 is

0.9	0.78333		-0.03123		0.00125
		0.010788		-0.00430		0.00016
1.1	0.89121		-0.03553		0.00141

and the values required for Bessel's formula (3.61) are

$$\mu f_{1/2} = \tfrac{1}{2}(f_0 + f_1) = \tfrac{1}{2}(0.78333 + 0.89121), \quad \delta f_{1/2} = 0.10788,$$
$$\mu\delta^2 f_{1/2} = \tfrac{1}{2}(\delta^2 f_0 + \delta^2 f_1) = \tfrac{1}{2}(-0.03123 - 0.03553), \quad \delta^3 f_{1/2} = 0.00430, \text{ etc.}$$

Since $x = x_0 + rh$ we have

$$0.98 = 0.9 + r(0.2),$$

giving $r = 0.4$. Substitution into Bessel's formula (3.61) gives

$$f(0.98) = 0.83727 - 0.01079 + 0.00401 - 0.00002 + 0.00003 - \cdots$$

and therefore $f(0.98) = 0.8305$ to four decimal places.

To estimate $f(1.08)$ we observe that $x = 1.08$ is closer to a tabular point than a mid-point. We therefore use Stirling's formula with $x_0 = 1.1$ because the associated formula pattern is symmetrical about 1.1, the closest tabular point to 1.08. Noting Stirling's formula pattern in Table 3.6 the required section of the numerical table in Example 3.5 is

		0.10788		-0.00430	
1.1	0.89121		-0.03553		0.00141
		0.07235		-0.00289	

and the values required for Stirling's formula are

$$f_0 = 0.89121, \quad \mu\delta f_0 = \tfrac{1}{2}(\delta f_{1/2} + \delta f_{-y_2} + \delta f_{-y_2}) = \tfrac{1}{2}(0.07235 + 0.10788),$$
$$\delta^2 f_0 = -0.03553, \quad \mu\delta^2 f_0 = \tfrac{1}{2}(\delta^3 f_{1/2} + \delta^3 f_{-1/2}) = \tfrac{1}{2}(-0.00430 - 0.00289), \text{ etc.}$$

Since $x = x_0 + rh$ we have

$$1.08 = 1.1 + r(0.2),$$

giving $r = -0.1$. Substitution into Stirling's formula (3.57) gives

$$f(1.08) = 0.89121 - 0.00901 - 0.00018 - 0.00006 - 6 \times 10^{-7} \cdots$$

and therefore $f(1.08) = 0.8820$ to four decimal places.

Algorithm 3.3. The Stirling formula to interpolate using a table with uniformly spaced points

The initial tabular point is $x0$ and the n tabular function values are held in $y[0], \ldots, y[n-1]$, with h the step length. The degree of the approximating polynomial is m and the required differences are held in Diff[0] & Diff[1]. The interpolated value of y at the point x is yx; terms are dealt with in pairs.

```
read(x0, h, n);
FOR i:= 0 TO n − 1 DO read(y[i]);
read(x, m);
j:= round((x − x0)/h);
r:= (x − x0)/h − j; rSqd:= sqr(r);
LoLim:= 1 − m/2; UpLim:= m/2;
IF (j − m/2 >= 0) AND (j + m/2 < n) THEN
  BEGIN
  FOR i:= LoLim TO UpLim DO Diff[i]:= y[j + i] − y[j + i − 1];
  Coeff:= r; yx:= y[j];
  FOR i:= 1 TO m/2 DO
    BEGIN
    yx:= yx + Coeff*(Diff[1] + Diff[0])/2;
    UpLim:= UpLim − 1;
    FOR j:= LoLim TO UpLim DO Diff[j]:= Diff[j + 1] − Diff[j];
    Coeff:= Coeff/(2*i);
    yx:= yx + r*Coeff*Diff[0];
    LoLim:= LoLim + 1;
    FOR j:= UpLim DOWNTO LoLim DO Diff[j]:= Diff[j] − Diff[j − 1];
    Coeff:= Coeff*(rSqd − sqr(i))/(2*i + 1)
    END;
  write(yx)
  END
ELSE write ('not enough function values').
```

Algorithm 3.4. The Bessel formula to interpolate using a table with uniformly spaced points

The initial tabular point is $x0$ and the n tabular function values are held in $y[0], \ldots,$ $y[n-1]$, with h the step length. The degree of the approximating polynomial is m and the required differences are held in Diff[0] & Diff[1]. The interpolated value of y at the point x is yx; terms are dealt with in pairs.

```
read(x0, h, n);
FOR i := 0 TO n − 1 DO read(y[i]);
read(x, m);
j := round((x − x0)/h);
r := (x − x0)/h − j; rMinus := r − 0.5;
LoLim := − m/2; UpLim := m/2;
IF (j − m/2 >= 0) AND (j + m/2 + 1 < n) THEN
    BEGIN
        FOR i := LoLim TO UpLim DO Diff[i] := y[j + i + 1] − y[j + i];
        Coeff := 1; yx := (y[j + 1] + y[j])/2;
    FOR i := 1 TO m/2 DO
        BEGIN
        yx := yx + rMinus*Coeff*Diff[0];
        LoLim := LoLim + 1;
        FOR j := UpLim DOWNTO LoLim DO Diff[j] := Diff[j] − Diff[j − 1];
        Coeff := (r + i − 1)*(r − i)*Coeff/(2*i);
        yx := yx + r*Coeff*(Diff[1] + Diff[0])/2;
        UpLim := UpLim − 1;
        FOR j := LoLim TO UpLim DO Diff[j] := Diff[j + 1] Diff[j];
        Coeff := Coeff/(2*i + 1)
        END;
    write(yx)
    END
ELSE write('not enough function values').
```

3.5 CHOICES OF INTERPOLATION POLYNOMIAL

With the exception of the Gauss forward and backward formulae, which were derived as an intermediate step in the derivation of Stirling's formula and Bessel's formula, we have demonstrated when and how to use the interpolation formulae that we have derived (Examples 3.5, 3.6, 3.7). Here we summarize this information and give some guidance on selection of an interpolation formula.

When the data are equally spaced with respect to x, always use a central difference interpolation formula unless the point of interpolation is very close to one end of the finite difference table (see below). This allows a choice between Stirling's formula and Bessel's formula which should be made with the aim of using data that are, as far as possible, symmetrically located about the point of interpolation. This consideration and the associated difference patterns show that Stirling's formula is the more suitable for interpolation close to a tabular point and Bessel's formula is the more suitable for interpolation close to a mid-point.

Although it is always preferable to use a central difference interpolation formula it is essential that the associated formula pattern lies within the triangular shape of the difference table. Consequently the Gregory–Newton forward formula must be used for interpolation close to the top of a difference table and the Gregory–Newton backward formula must be used for interpolation close to the bottom of a difference table. The difference patterns associated with other interpolation formulae quickly run out of the table.

EXERCISES 3

1. Form a finite difference table for the function $f(x)$ that takes the values 2.31700, 2.42542, 2.52208, 2.60944, 2.68924, 2.76275 for $x = 1.0(0.2)2.0$. Choosing $x_0 = 1.4$, extract values from the table for Δf_2, ∇f_2, $\nabla^3 f_1$, $\nabla^2 f_{-1}$, $\Delta^4 f_{-2}$, $\delta f_{3/2}$, $\delta^2 f_{1/2}$, $\mu\delta^2 f_{1/2}$, $\delta^2 f_{-1}$, $\delta^3 f_{1/2}$, and $\mu\delta^3 f_1$, if they exist.

2. From the definitions of the operators involved show that

   (i) $\nabla^3 f_k = f_k - 3f_{k-1} + 3f_{k-2} - f_{k-3}$
   (ii) $\delta^3 f_k = f_{k+3/2} - 3f_{k+1/2} + 3f_{k-1/2} - f_{k-3/2}$.

   Observe the values of the suffices on the right-hand side of each equation relative to k (also see Example 3.2). Use these expressions to determine $\nabla^3 f_1$ and $\delta^3 f_{1/2}$ for the function specified numerically in Exercise 1. Check that your answer agrees with those obtained directly from the difference table in Exercise 1.

3. Determine $\Delta f(x)$ for the following functions:

   (i) $f(x) = 2x^3 - 3x^2 + x - 2$,
   (ii) $f(x) = \log ax$, $a = $ constant,
   (iii) $f(x) = 1/x^2$.

4. Prove the operator identities

   (i) $\nabla = 1 - E^{-1}$
   (ii) $\delta = E^{1/2} - E^{-1/2}$
   (iii) $\mu = \frac{1}{2}(E^{1/2} + E^{-1/2})$
   (iv) $E = 1 + \dfrac{\delta^2}{2} + \delta\left\{1 + \dfrac{\delta^2}{4}\right\}^{1/2}$.

   (Hint: (iv) follows from (ii), but a plus or minus sign appears in front of the square root in (iv). Checking (iv) with a simple test function such as $f(x) = x$ shows that the minus sign should be discarded. The algebraic manipulation (squaring) introduces this incorrect solution.)

5. Show that

   (i) $\Delta \tan^{-1}(x) = \tan^{-1}\left[\dfrac{h}{x^2 + hx + 1}\right]$,

(ii) $\Delta \sin^{-1}(x) = \sin^{-1}[(x+h)\sqrt{(1-x^2)} - x\sqrt{\{1-(x+h)^2\}}]$ for sufficient small h,

(iii) $\Delta \sin ax = 2 \sin \left\{ \dfrac{ah}{2} \right\} \cos \left\{ a \left(x + \dfrac{h}{2} \right) \right\}$,

(iv) $\Delta \cos ax = -2 \sin \left\{ \dfrac{ah}{2} \right\} \sin \left\{ a \left(x + \dfrac{h}{2} \right) \right\}$.

6. Show that

(i) $\Delta^m f_k = \delta^m E^{m/2} f_k = \nabla^m E^m f_k$,

(ii) $\Delta \left(\dfrac{f_k}{g_k} \right) = \dfrac{g_k \Delta f_k - f_k \Delta g_k}{g_k g_{k+1}}$,

(iii) $\delta \left(\dfrac{1}{f_{k+1/2}} \right) = -\dfrac{\delta f_{k+1/2}}{f_k f_{k+1}}$.

7. Show that

(i) $\delta \sin^2 ax = \sin ah \sin 2ax$,

(ii) $\delta \tan ax = \dfrac{\sin ah}{\cos a(x - h/2) \cos a(x + h/2)}$,

(iii) $\delta^r \left(\dfrac{1}{x} \right) = \dfrac{(-1)^r r! \, h^r}{(x - rh/2) \cdots (x - h/2)x(x + h/2) \cdots (x + rh/2)}$,

(iv) $\delta^m \sin ax = \left(2 \sin \dfrac{ah}{2} \right)^m \sin \left\{ ax + \dfrac{m\pi}{2} \right\}$,

(v) $\delta^m \cos ax = \left(2 \sin \dfrac{ah}{2} \right)^m \cos \left\{ ax + \dfrac{m\pi}{2} \right\}$,

(vi) $\delta^m \cos^2 ax = \frac{1}{2}(2 \sin ah)^m \cos \left\{ 2ax + \dfrac{m\pi}{2} \right\}$.

8. Derive equations (3.24) and (3.25) from equation (3.23) and Exercise 6(i).

9. Determine whether the functions represented by the following data can be approximated by polynomials over the specified ranges:

(i) $f(x)$ takes the values 1.0000, 1.2840, 2.7183, 9.4877, 54.5982 for $x = 0.0(0.5)2.0$.
(ii) $f(x)$ takes the values 1.4142, 1.5811, 1.7321, 1.8708, 2.0000 for $x = 2.0(0.5)4.0$.

10. Estimate the values of $f(1.15)$ and $f(1.98)$ for the function specified numerically in Exercise 1 and state the accuracy of your answer.

11. A function takes the values 1.0000, 1.0960, 1.0480, 0.9520, 0.9040, 1.00000 for $x = 0.0(0.2)1.0$. Show that a cubic can represent this set of data over the range considered. Determine this cubic in terms of x.

12. Show that
$$p_3(x) = f_1 + r\Delta f_1 + \frac{r(r-1)}{2!}\delta^2 f_1 + \frac{r(r-1)(r-2)}{3!}\delta^3 f_1,$$

where $x = x_1 + rh$, and

$$q_3(x) = f_4 + s\nabla f_4 + \frac{s(s+1)}{2!}\nabla^2 f_4 + \frac{s(s+1)(s+2)}{3!}\nabla^3 f_4,$$

where $x = x_4 + sh$, are alternative forms of the cubic polynomial passing through (x_1, f_1), (x_2, f_2), (x_3, f_3) and (x_4, f_4).

13. The function $f(x)$ takes the values 1.00000, 1.04603, 1.09417, 1.14454, 1.29722, 1.25232 for $x = 0.0(0.1)0.5$. Estimate values for $f(0.05)$ and $f(0.47)$.

14. Express each of

$$f_1(x) = f[0] + xf[0, 0.5] + x(x - 0.5)f[0, 0.5, 1.5]$$
$$+ x(x - 0.5)(x - 1.5)f[0, 0.5, 1.5, 2.1]$$

and

$$f_2(x) = f[1.5] + (x - 1.5)f[1.5, 0] + (x - 1.5)xf[1.5, 0, 2.1]$$
$$+ (x - 1.5)x(x - 2.1)f[1.5, 0, 2.1, 0.5]$$

in the form $ax^3 + bx^2 + cx + d$ given that $f(0) = 1$, $f(0.5) = 1.375$, $f(1.5) = 0.625$, $f(2.1) = 1.231$. Verify that both are alternative forms of the cubic passing through $(0, 1)$, $(0.5, 1.375)$, $(1.5, 0.625)$ and $(2.1, 1.231)$.

15. From the data $(0.25, 0.77110)$, $(0.37, 1.17941)$, $(0.42, 1.36233)$, $(0.52, 1.75788)$, $(0.60, 2.10946)$, $(0.75, 2.87928)$ for $(x, f(x))$, estimate values for $f(0.3)$, $f(0.5)$ and $f(0.7)$, giving the accuracy of your answers.

16. Use mathematical induction with respect to k to prove equation (3.34).

17. The function $f(x)$ takes the values 0.25000, 0.33693, 0.46769, 0.65266, 0.91103, 1.27388 for $x = 0.0(0.2)1.0$. Estimate values for $f(0.37)$ and $f(0.52)$, stating the accuracy of your answers.

18. The function $J(x)$ takes the values 0.0000, 0.0995, 0.1960, 0.2867, 0.3688, 0.4401, 0.4983, 0.5419 for $x = 0.0(0.2)1.4$. Use the most appropriate interpolation formula to estimate $J(0.1)$, $J(0.72)$, $J(0.77)$ and $J(1.32)$, stating the accuracy of your answers.

19. Derive the Gauss backward formula, equation (3.54), from Newton's divided difference formula.

20. (i) Show that Stirling's formula (3.57), when terminated after the second-order difference term, is a form of the quadratic passing through (x_{-1}, f_{-1}), (x_0, f_0) and (x_1, f_1) but that Bessel's formula (3.61) is not.
    (ii) Show that Bessel's formula (3.61), when terminated after the third-order difference term, is a form of the cubic passing through (x_{-1}, f_{-1}), (x_0, f_0), (x_1, f_1) and (x_2, f_2) but that Stirling's formula (3.57) is not.

4

Numerical Integration

Two situations can arise in which we need to obtain a numerical approximation to the definite integral

$$I = \int_a^b f(x)\,dx. \tag{4.1}$$

These are

(i) the explicit form of the integrand $f(x)$ is not known but is specified numerically by the data points (x_i, f_i) for $i = 0(1)n$, where $x_0 = a$ and $x_n = b$, and

(ii) the explicit form of the integrand is known but is too complicated to be integrated analytically.

However, when the integrand is known explicitly, a set of data (x_i, f_i) for $i = 0(1)n$ can always be generated. Thus we develop methods for solving problems of type (i), and the same methods can be applied to problems of type (ii). We limit our discussion to the situation when the data are equally spaced with respect to x, i.e. $x_{i+1} = x_i + h$.

When the integrand is specified by the data points (x_i, f_i), $i = 0(1)n$, the problem of deriving a formula that approximates the integral (4.1) amounts to finding an approximate continuous representation of the integrand over the range $[a, b]$. This can be achieved by fitting interpolation polynomials to the data (x_i, f_i), $i = 0(1)n$, as discussed in the previous chapter. Consequently the elementary integration of interpolation polynomials leads to formulae that approximate the definite integral (4.1).

Given a set of equally spaced data, (x_i, f_i), $i = 0(1)n$, there are many ways in which interpolation polynomials can be used approximate the integrand over the entire range. One method is to use a polynomial of degree n that passes through the $n + 1$ data points (x_i, f_i). With this approach the integrand is approximated by a function that is smooth on $[x_0, x_n]$, i.e. the approximating function has a continuous derivative on

63

$[x_0, x_n]$. However, if n is large, this leads to a complicated formula. Alternatively the interval $[x_0, x_n]$ can be divided into sub-intervals, and the integrand can be approximated on each sub-interval by a different interpolation polynomial. For example, we could divide $[x_0, x_n]$ into the sub-intervals

$$[x_0, x_2], [x_2, x_4], [x_4, x_6], \ldots.$$

For each of these sub-intervals $[x_i, x_{i+2}]$, $i = 0(2)n - 2$, there are three data points, (x_i, f_i), (x_{i+1}, f_{i+1}) and (x_{i+2}, f_{i+2}). Consequently the integrand can be approximated by a quadratic interpolation polynomial on each sub-interval. With this approach the approximation is only smooth within each interval $[x_i, x_{i+2}]$ since there is a discontinuity in the slope at each of the points $x_2, x_4, \ldots, x_{n-2}$, where the quadratics from adjoining sub-intervals meet (see Fig. 4.1). Such a function that is continuous on an interval $[a, b]$ and smooth on $[a, b]$ except at a finite number of points is described as piecewise-smooth.

There are many ways in which the interval $[x_0, x_n]$ can be divided into sub-intervals leading to many piecewise-smooth polynomial approximations of the integrand over $[x_0, x_n]$. However, there are two distinct approaches to obtaining formulae for approximating the integral (4.1) depending on whether smooth or piecewise-smooth approximations are used for the integrand. Consequently two classes of integration formula arise. The integration formulae that arise from smooth polynomial approximations to the integrand are described as Newton–Cotes formulae. Those that result from piecewise-smooth polynomial approximations are described as composite formulae. Later it will become evident that composite formulae are obtained by adding appropriate Newton–Cotes formulae.

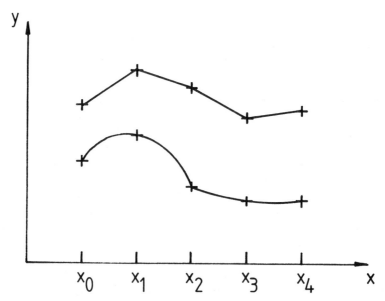

Fig. 4.1. Piecewise-linear and piecewise-quadratic approximations to two sets of data specified in the form (x_i, f_i), $i = 0(1)4$.

4.1 APPROXIMATION OF THE INTEGRAND

We describe a systematic method for deriving integration formulae from piecewise-smooth approximations to the integrand. Each data point (x_i, f_i) can be represented by a point P_i in the xy-plane. Then the total data set is $\{P_i : i = 0(n)n\}$ and the index i orders the points according to the corresponding x-value since $x_{i+1} = x_i + h$. If we divide $[x_0, x_n]$ into sub-intervals $[x_i, x_{i+m}]$ of equal length, each sub-interval has the associated set of $m + 1$ data points $\{P_i, P_{i+1}, \ldots, P_{i+m}\}$. Taking $m = 1$ gives the simplest division of this form,

$$[x_0, x_1], [x_1, x_2], [x_2, x_3], \ldots,$$

and, correspondingly, the data set is divided into pairs of adjacent points

$$(P_0, P_1), (P_1, P_2), (P_2, P_3), \ldots.$$

The corresponding piecewise-smooth polynomial approximation is obtained by joining each pair of points with a straight line (Fig. 4.1). For $m = 2$ the division $[x_i, x_{i+2}]$, $i = 0(2)n - 2$, noted in the previous section, is obtained and the corresponding division of data is

$$(P_0, P_1, P_2), (P_2, P_3, P_4), (P_4, P_5, P_6), \ldots,$$

i.e. the data are divided into subsets of adjacent points taken three at a time and a quadratic can be passed through each subset of three points. This procedure can be extended in the obvious way with divisions of $[x_0, x_n]$ leading to subsets of adjacent data points taken four at time, five at time, and soon. Such divisions are only possible when n (there are $n + 1$ data points) is divisible by m, where $m + 1$ is the number of points in each subset. (Consider $[x_0, x_{10}]$ as an example. Here $n = 10$ and there are $n + 1 = 11$ data points. Since $n = 10$ the possible divisions are given by $m = 1, 2, 5$. The corresponding divisions of data are (P_i, P_{i+1}) for $i = 0(1)9$, (P_i, P_{i+1}, P_{i+2}) for $i = 0(2)8$ and $(P_i, P_{i+1}, P_{i+2}, P_{i+3}, P_{i+4}, P_{i+5})$ for $i = 0, 5$. There are $m + 1 = 2, 3, 6$ points in each set respectively.)

The method described above leads to successive approximations to the integrand $f(x)$ that are piecewise-linear, piecewise-quadratic, piecewise-cubic, and so on, depending on the number of points in each subset of $\{P_i : i = 0(1)n\}$. The approximations to $f(x)$ are interpolation polynomials that pass through the data points in each subset. In Chapter 3 we showed that the interpolation polynomial through a given set of points can be expressed in alternative forms. Many of these forms can be used to derive integration formulae. Here we use the Gregory–Newton forward interpolation formula (equation (3.30)) because of its relative simplicity and to give a unified approach to the derivations of the integration formulae. Recall that the first two terms of this Gregory–Newton formula give the linear polynomial passing through P_i and P_{i+1}, the first three terms yield the quadratic passing through P_i, P_{i+1} and P_{i+2}, and so on. In general, the interpolation polynomial passing through the points in the subset $\{P_i, P_{i+1}, \ldots, P_{i+m}\}$ has degree m and can be obtained by terminating the Gregory–Newton forward formula after the term containing the mth-order difference.

A division $\{P_i, P_{i+1}, \ldots, P_{i+m}\}$ of $\{P_i : i = 0(1)n\}$ leads to both a Newton–Cotes integration formula and a composite integration formula for each value of m. Provided *n* is divisible by *m* we may write

$$\int_{x_0}^{x_n} f(x)\,dx = \sum_{i=0(m)n-m} \int_{x_i}^{x_{i+m}} f(x)\,dx \tag{4.2}$$

(see (4.3), (4.19), for example) and on each interval $[x_i, x_{i+m}]$ the integrand $f(x)$ can be approximated by the polynomial passing through all points in the subset $\{P_i, P_{i+1}, \ldots, P_{i+m}\}$. This smooth approximation leads to a Newton–Cotes formula for

$$\int_{x_i}^{x_{i+m}} f(x)\,dx.$$

By taking $m = 1, 2, 3, \ldots$ a set of increasingly complex Newton–Cotes formulae is obtained. Each Newton–Cotes formula that is obtained for a particular value of *m* can be substituted into (4.2) to obtain a composite formula for

$$\int_{x_0}^{x_n} f(x)\,dx.$$

Each Newton–Cotes formula gives an approximation to the integral of $f(x)$ on a sub-interval of $[x_0, x_n]$ and there is an associated error. This error arises because $f(x)$ is approximated on a sub-interval by truncating the Gregory–Newton formula. Consequently the error can be examined by looking at the remaining terms of the Gregory–Newton formula. The truncation error associated with each Newton–Cotes formula is a local truncation error, i.e. local to the sub-interval for which the Newton–Cotes formula has been derived. The error associated with a composite integration formula will be the sum of all the errors associated with the Newton–Cotes formulae from which it is composed. This is usually referred to as global truncation error. These ideas are demonstrated below.

4.2 SOME WELL-KNOWN INTEGRATION FORMULAE

4.2.1 The Trapezium rule

The Trapezium rule can be derived by dividing $[x_0, x_n]$ into the sub-intervals $[x_i, x_{i+1}]$ for $i = 0(1)n - 1$ and splitting the integral (4.1) accordingly:

$$\int_{x_0}^{x_n} f(x)\,dx = \int_{x_0}^{x_1} f(x)\,dx + \int_{x_1}^{x_2} f(x)\,dx + \cdots + \int_{x_{n-1}}^{x_n} f(x)\,dx. \tag{4.3}$$

Then, on a sub-interval $[x_i, x_{i+1}]$, the integrand $f(x)$ is approximated by the straight line joining P_i and P_{i+1}. If $x = x_i + rh$ the Gregory–Newton forward expansion for $f(x)$ is (see equation (3.30))

$$f(x) = f_i + r\Delta f_i + \frac{r(r-1)}{2!}\Delta^2 f_i + \frac{r(r-1)(r-2)}{3!}\Delta^3 f_i + \cdots \tag{4.4}$$

and the first two terms of the expansion give the straight line joining P_i and P_{i+1}. Thus to derive the Trapezium rule we make the approximation to the integrand

$$f(x) \simeq f_i + r\Delta f_i, \quad x_i \leqslant x \leqslant x_{i+1}, \quad 0 \leqslant r \leqslant 1. \tag{4.5}$$

That is, for the interval $[x_i, x_{i+1}]$ we take

$$\int_{x_i}^{x_{i+1}} f(x)\,dx = \int_0^1 f(x_i + rh)\,h\,dr$$

$$= \int_0^1 \{f_i + r\Delta f_i\}\,h\,dr + \varepsilon_i, \tag{4.6}$$

where ε_i is the local truncation error to the integration formula on $[x_i, x_{i+1}]$ associated with the linear approximation (4.5). From equations (4.4) and (4.6) we see that this local truncation error is given by

$$\varepsilon_i = \int_0^1 \left\{ \frac{r(r-1)}{2!}\Delta^2 f_i + \frac{r(r-1)(r-2)}{3!}\Delta^3 f_i + \cdots \right\} h\,dr. \tag{4.7}$$

The limits for r in (4.6) and (4.7) have been obtained from the linear relation $x = x_i + rh$, i.e. $r = 0$ when $x = x_i$ and $r = 1$ when $x = x_{i+1}$. This relation also leads to $dx = (dx/dr)\,dr = h\,dr$.

Integrating (4.6) yields

$$\int_{x_i}^{x_{i+1}} f(x)\,dx = h\left[rf_i + \frac{r^2}{2}\Delta f_i \right]_0^1 + \varepsilon_i$$

$$= h[f_i + \tfrac{1}{2}\Delta f_i] + \varepsilon_i$$

$$= \frac{h}{2}(f_i + f_{i+1}) + \varepsilon_i \tag{4.8}$$

since $\Delta f_i = f_{i+1} - f$.

We obtain an expression for the error ε_i using an approach that lacks mathematical rigour but leads to the correct form. We consider the leading contribution to the error in (4.7). Thus, integrating only the first term of (4.7) we obtain

$$\varepsilon_i = -\frac{h}{12}\Delta^2 f_i + \cdots. \tag{4.9}$$

It is normal practice to express the error in terms of derivatives rather than differences. This is more useful because differences depend on the spacing h used in a particular problem. This convention allows us to classify and compare the error terms associated with different numerical integration formulae. To express (4.9) in terms of derivatives we relate the differential operater $D = d/dx$ to the forward difference operator Δ. Using the definition of Δ and a Taylor expansion we have

$$\Delta f(x) = f(x + h) - f(x)$$

$$= \{ f(x) + hf'(x) + \frac{h^2}{2}f''(x) + \cdots \} - f(x)$$

$$= \{ hD + \frac{h^2}{2!}D^2 + \cdots \} f(x).$$

Thus the operator in curly brackets must be equivalent to Δ, i.e.

$$\Delta = hD + \frac{h^2 D^2}{2!} + \cdots.$$

In general we will use this result in the form

$$\Delta^s = \left\{ hD + \frac{h^2 D^2}{2!} + \cdots \right\}^s$$

$$= h^s D^s + \cdots \tag{4.10}$$

for different values of s and consider the leading term only. Using (4.10), with $s = 2$, equation (4.9) can be written

$$\varepsilon_i = -\frac{h^3}{12} f''(x_i) + \cdots. \tag{4.11}$$

Thus our non-rigorous approach shows that the leading contribution to the error term on $[x_i, x_{i+1}]$ is $-h^3 f''(x_i)/12$. Using a more rigorous mathematical derivation it can be shown that the error on $[x_i, x_{i+1}]$ is given exactly by

$$\varepsilon_i = -\frac{h^3}{12} f''(\xi_i), \tag{4.12}$$

where ξ_i lies in the interval $[x_i, x_{i+1}]$. Notice that this has the same form as the leading contribution that we derived, but the value of x has changed from $x = x_i$ to $x = \xi_i$. Combining equations (4.8) and (4.12) the Newton–Cotes form of the Trapezium rule, with error term, applied to the interval $[x_i, x_{i+1}]$ is

$$\int_{x_i}^{x_{i+1}} f(x) \, dx = \frac{h}{2} (f_i + f_{i+1}) - \frac{h^3}{12} f''(\xi_i). \tag{4.13}$$

Similarly, the Newton–Cotes form of the Trapezium rule for an integral over $[a, b]$ is

$$\int_a^b f(x) \, dx = \frac{h}{2} (f(a) + f(b)) - \frac{h^3}{12} f''(\xi), \tag{4.14}$$

where $h = b - a$ (recall $h = x_{i+1} - x_i$ in (4.13)) and $a \leqslant \xi \leqslant b$.

The composite form of the Trapezium rule for estimating the definite integral (4.1) can be obtained by substituting (4.13) into (4.3):

$$\int_{x_0}^{x_n} f(x) \, dx = \sum_{i=0}^{n-1} \int_{x_i}^{x_{i+1}} f(x) \, dx$$

$$= \sum_{i=0}^{n-1} \left\{ \frac{h}{2} (f_i + f_{i+1}) - \frac{h^3}{12} f''(\xi_i) \right\}.$$

Thus the composite form of the Trapezium rule is

$$\int_{x_0}^{x_n} f(x) \, dx = h \left\{ \frac{f_0}{2} + f_1 + f_2 + \cdots + f_{n-1} + \frac{f_n}{2} \right\} + \varepsilon_G \tag{4.15}$$

where ε_G is the global error and is given by

$$\varepsilon_G = -\frac{h^3}{12} \{ f''(\xi_0) + f''(\xi_1) + \cdots + f''(\xi_{n-1}) \}. \tag{4.16}$$

The global error is not particularly useful in this form. However, we can obtain an upper bound for ε_G from (4.16). Since

$$|f''(\xi_i)| \leqslant \max_{a \leqslant x \leqslant b} \{|f''(x)|\}$$

for $i = 0(1)n - 1$, then summing gives

$$\sum_{i=0}^{n-1} |f''(\xi_i)| \leqslant n \max_{a \leqslant x \leqslant b} \{|f''(x)|\}$$

and from (4.16) we obtain

$$|\varepsilon_G| \leqslant \frac{nh^3}{12} \max_{a \leqslant x \leqslant b} \{|f''(x)|\}. \tag{4.17}$$

Observing that $nh = b - a$ the inequality (4.17) can also be written in the form

$$|\varepsilon_G| \leqslant (b - a)\frac{h^2}{12} \max_{a \leqslant x \leqslant b} \{|f''(x)|\}. \tag{4.18}$$

The right-hand sides of (4.17) and (4.18) give upper bounds for the global error associated with the composite form of the Trapezium rule.

As expected, both the local truncation error and the global error depend on the step size h, and smaller values of h head to smaller errors. We can make use of this observation to estimate an integral to a required accuracy. If we calculate a sequence of estimates $T(h_0)$, $T(h_1)$, $T(h_2)$, ... for the integral (4.1) using the Trapezium rule with step lengths h_0, $h_1 = h_0/2$, $h_2 = h_1/2$, and so on, the sequence will converge to the exact value of the integral. Thus when we need to estimate an integral to four decimal places, for example, we calculate $T(h_0)$, $T(h_1)$, $T(h_2)$, and so on until two successive members of the sequence agree when rounded to four decimal places. This is demonstrated in the following example.

Example 4.1 Use the Trapezium rule to estimate

$$\int_0^{0.8} e^{x^2}\, dx$$

to two decimal places.

Since two-decimal-place accuracy is asked for, intermediate working is carried out to the three decimal places. An initial estimate to the integral can be obtained using the step length $h_0 = 0.4$. This leads to the estimate ((4.15) with $n = 2$, $h = 0.4$)

$$I(0.4) = 0.4\left(\frac{f(0)}{2} + f(0.4) + \frac{f(0.8)}{2}\right),$$

$$= 1.049$$

since $f(x) = e^{x^2}$. Halving the step length to $h_1 = 0.2$ gives the second estimate ((4.15) with $n = 4$)

$$I(0.2) = 0.2\left(\frac{f(0)}{2} + f(0.2) + f(0.4) + f(0.6) + \frac{f(0.8)}{2}\right)$$

$$= 1.019.$$

Clearly further estimates are required, and using $h_2 = 0.1$ we obtain

$$I(0.1) = 0.1\left(\frac{f(0)}{2} + f(0.1) + f(0.2) + f(0.3) + f(0.4) + f(0.5)\right.$$

$$\left. + f(0.6) + f(0.7) + \frac{f(0.8)}{2}\right)$$

$$= 1.012.$$

Finally with $h_3 = 0.05$ we find that $I(0.05) = 1.010$, which agrees with $I(0.1)$ when rounded to two decimal places. In summary, the required integral takes the value 1.01 to two decimal places.

It is evident that this process of repeatedly halving the step length and obtaining several Trapezium rule estimates can be laborious. However there is an alternative which is usually more efficient. The upper bound for the global truncation error in (4.18) can be used to estimate a step length h that will yield a required accuracy. Thus, if four-decimal-place accuracy is required, choose h such that

$$(b - a)\frac{h^2}{12} \max_{a \leqslant x \leqslant b} \{|f''(x)|\} < 0.00005.$$

This is only possible when the integrand is known. This technique may estimate a value for h that is smaller than is necessary to obtain the required accuracy since it is based on an upper bound for the global truncation error. However it is usually more efficient than the method used in the above example. This approach is demonstrated at the end of Example 4.2.

Algorithm 4.1 provides a Trapezium rule estimate $T(h)$ to the definite integral (4.1). The reader is left to consider how to modify this algorithm to calculate $T(h)$, $T(h/2)$, $T(h/4),\ldots$ until two successive members of this sequence differ by less than ε. Then setting $\varepsilon = 0.000005$, for example, would give accuracy to four decimal places.

Algorithm 4.1. The Trapezium rule

The interval of integration is $[a, b]$ and the number of intervals is n; the function f is to be specified.

```
read(a, b, n);
h:= (b − a)/n;
Sum:= (f(a) + f(b))/2
FOR i:= 1 TO n − 1 DO Sum:= Sum + f(a + i*h);
IntegVal:= h*Sum;
write(IntegVal).
```

4.2.2 Simpson's rule

Having derived the integration formula associated with a piecewise-linear approximation to the integrand we now derive a formula associated with a piecewise-quadratic approximation to the integrand. Thus we divide the interval $[x_0, x_n]$ into the sub-intervals $[x_i, x_{i+2}]$ for $i = 0(2)n - 2$ and express the integral in the form

$$\int_{x_0}^{x_n} f(x)\,dx = \sum_{i=0(2)n-2} \int_{x_i}^{x_{i+2}} f(x)\,dx. \tag{4.19}$$

On a sub-interval $[x_i, x_{i+2}]$ the integrand $f(x)$ is approximated by the quadratic passing through P_i, P_{i+1} and P_{i+2}. Taking $x = x_i + rh$, the first three terms of the Gregory–Newton forward formula (4.4) give the required quadratic:

$$f(x) \simeq f_i + r\Delta f_i + \frac{r(r-1)}{2}\Delta^2 f_i, \; x_i \leqslant x \leqslant x_{i+2}, 0 \leqslant r \leqslant 2. \tag{4.20}$$

Thus the contribution to (4.19) over $[x_i, x_{i+2}]$ is

$$\int_{x_i}^{x_{i+2}} f(x)\,dx = \int_0^2 f(x_i + rh)h\,dr$$

$$= \int_0^2 \left\{ f_i + r\Delta f_i + \frac{(r^2 - r)}{2}\Delta^2 f_i \right\} h\,dr + \varepsilon_i. \tag{4.21}$$

where ε_i is the local truncation error to the integration formula on $[x_i, x_{i+2}]$ associated with the quadratic approximation (4.20). From equations (4.20), (4.4) we see that this local truncation error is given by

$$\varepsilon_i = \int_0^2 \left\{ \frac{r(r-1)(r-2)}{3!}\Delta^3 f_i + \frac{r(r-1)(r-2)(r-3)}{4!}\Delta^4 f_i + \cdots \right\} h\,dr. \tag{4.22}$$

The limits for r in (4.21) and (4.22) follow from the linear relation $x = x_i + rh$, i.e. $r = 0$ when $x = x_i$ and $r = 2$ when $x = x_{i+2}$.

Integrating (4.21) yields

$$\int_{x_i}^{x_{i+2}} f(x)\,dx = h[2f_i + 2\Delta f_i + \tfrac{1}{3}\Delta^2 f_i] + \varepsilon_i$$

and substituting $\Delta f_i = f_{i+1} - f_i$, $\Delta^2 f_i = f_{i+2} - 2f_{i+1} + f_i$ leads to

$$\int_{x_i}^{x_{i+2}} f(x)\,dx = \frac{h}{3}[f_i + 4f_{i+1} + f_{i+2}] + \varepsilon_i. \tag{4.23}$$

As with the Trapezium rule we only determine the leading contribution to the error in (4.22) and note that a more rigorous mathematical approach shows that the exact error is of a similar form to this leading contribution. Integrating equation (4.22) we find that the coefficient of $\Delta^3 f_i$ is zero so that the leading contribution comes from the term involving $\Delta^4 f_i$. Carrying out the elementary integration leads to

$$\varepsilon_i = -\frac{h}{90}\Delta^4 f_i + \cdots \tag{4.24}$$

and using (4.11) with $s = 4$ gives

$$\varepsilon_i = -\frac{h^5}{90}f^{(iv)}(x_i) + \cdots. \tag{4.25}$$

We have determined the leading contribution to the error on $[x_i, x_{i+2}]$. Using a more rigorous mathematical approach it can be shown that the error on $[x_i, x_{i+2}]$ is given exactly by

$$\varepsilon_i = -\frac{h^5}{90} f^{(iv)}(\xi_i),\tag{4.26}$$

where ξ_i lies in the interval $[x_i, x_{i+2}]$. Combining equations (4.23) and (4.26), the Newton–Cotes form of Simpson's rule, with error term, applied to the interval $[x_i, x_{i+2}]$ is

$$\int_{x_i}^{x_{i+2}} f(x)\,dx = \frac{h}{3}[f_i + 4f_{i+1} + f_{i+2}] - \frac{h^5}{90} f^{iv}(\xi_i).\tag{4.27}$$

A first approximation to the integral (4.1) could be obtained by applying the Newton–Cotes form of Simpson's rule to the interval $[a, b]$. This gives

$$\int_a^b f(x)\,dx \simeq \frac{h}{3}\left[f(a) + 4f\left(\frac{a+b}{2}\right) + f(b) \right].\tag{4.28}$$

The composite form of Simpson's rule for estimating the definite integral (4.1) can be obtained by substituting (4.27) into (4.19):

$$\int_{x_0}^{x_n} f(x)\,dx = \sum_{i=0(2)n-2} \left\{ \frac{h}{3}[f_i + 4f_{i+1} + f_{i+2}] - \frac{h^5}{90} f^{iv}(\xi_i) \right\}.$$

This can be written

$$\int_{x_0}^{x_n} f(x)\,dx = \frac{h}{3}\{f_0 + 4f_1 + 2f_2 + 4f_3 + 2f_4 + \cdots + 2f_{n-2} + 4f_{n-1} + f_n\} + \varepsilon_G\tag{4.29}$$

where ε_G is the global error and is given by

$$\varepsilon_G = -\frac{h^5}{90}\{f^{(iv)}(\xi_0) + f^{(iv)}(\xi_2) + \cdots + f^{(iv)}(\xi_{n-2})\}.\tag{4.30}$$

To remember the composite form of Simpson's rule it is useful to observe the pattern of coefficients $142424\cdots241$. Alternatively, observe that four is the coefficient of all function values with odd suffices in the composite form of Simpson's rule and that two is the coefficient of the function values with even suffices. The first and last function values, f_0 and f_n, are exceptions.

Simpson's composite formula can only be applied when $n + 1$, the number of data points, is odd. When $n + 1$ is odd there are an even number of sub-intervals $[x_i, x_{i+1}]$ which can be paired off into the sub-intervals $[x_i, x_{i+2}]$. The division in (4.19) is then possible but is not possible when $n + 1$ is even. This places a restriction on the use of Simpson's composite formula when the integrand is only known in the form of a discrete set of numerical data. If the integrand is known analytically an odd number of equally spaced data points can always be generated.

Although the form of the global truncation error in (4.30) is of little use in practice, it does lead to a useful upper bound. Since

$$|f^{(iv)}(\xi_i)| \leqslant \max_{a \leqslant x \leqslant b} \{|f^{(iv)}(x)|\}$$

for $i = 0(2)n - 2$, then

$$\sum_{i=0(2)n-2} |f^{(iv)}(\xi_i)| \leqslant \frac{n}{2} \max_{a \leqslant x \leqslant b} \{|f^{(iv)}(x)|\}$$

and from (4.30) we obtain

$$|\varepsilon_G| \leqslant \frac{nh^5}{180} \max_{a \leqslant x \leqslant b} \{|f^{(iv)}(x)|\}. \tag{4.31a}$$

Observing that $nh = b - a$ this inequality can also be written

$$|\varepsilon_G| \leqslant (b - a) \frac{h^4}{180} \max_{a \leqslant x \leqslant b} \{|f^{(iv)}(x)|\}. \tag{4.31b}$$

The right-hand sides of (4.31a) and (4.31b) give upper bounds for the global error associated with Simpson's composite formula.

Example 4.2 By repeated application of Simpson's rule, use

$$\tan^{-1} x = \int_0^x \frac{1}{1 + t^2} \, dt$$

to determine $\tan^{-1}(1)$ to six decimal places. Determine an upper bound for the error in the final answer.

We have to obtain an estimate for

$$\int_0^1 \frac{1}{1 + t^2} \, dt$$

accurate to six decimal places. Consequently all intermediate working will be to seven decimal places. We will use the notation $S(h)$ to denote the estimate obtained using Simpson's rule with step length h. We calculate the approximations $S(1/2)$, $S(1/4)$, $S(1/8), \ldots$ until two successive approximations agree when rounded to six decimal places. Using equation (4.29) with $h = 1/2$ ($n = 2$) we have

$$S(1/2) = \tfrac{1}{6} \{ f(0) + 4f(1/2) + f(1) \}$$

and since $f(t) = 1/(1 + t^2)$ this yields

$$S(1/2) = 0.7833333.$$

Using the step length $h = 1/4$, Simpson's formula yields

$$S(1/4) = \tfrac{1}{12} \{ f(0) + 4f(1/4) + 2f(1/2) + 4f(3/4) + f(1) \}$$
$$= 0.7853922.$$

Since $S(1/2)$ and $S(1/4)$ do not agree to six decimal places we apply Simpson's composite formula with $h = 1/8$:

$$S(1/8) = \tfrac{1}{24} \{ f(0) + 4f(1/8) + 2f(1/4) + 4f(3/8) + 2f(1/2)$$
$$+ 4f(5/8) + 2f(3/4) + 4f(7/8) + f(1) \}$$
$$= 0.7853981.$$

At this stage, four-decimal-place accuracy has been reached between $S(1/4)$ and $S(1/8)$, but further calculation is required to obtain six-decimal accuracy. However we find that

$$S(1/6) = 0.7853982,$$

which agrees with $S(1/8)$ when both are rounded to six decimal places. Thus we have found that

$$\tan^{-1}(1) = 0.785398.$$

We calculate the right-hand side of (4.31) to determine an upper bound for the error in the estimate $S(1/16)$. Thus we require the maximum value of the fourth derivative of $f(t) = (1+t^2)^{-1}$ on $[0, 1]$. By repeated differentiation we obtain

$$f^{(iv)}(t) = \frac{24}{(1+t^2)^5}(5t^4 - 10t^2 + 1).$$

To determine the maximum value of $f^{(iv)}(t)$ on $[0, 1]$ we examine the stationary values on $[0, 1]$. Differentiating yields

$$f^{(v)}(t) = -\frac{240t}{(1+t^2)^6}(3t^2 - 1)(t^2 - 3)$$

and therefore $|f^{(iv)}(t)|$ will attain its maximum value at any of $t = 0, 1/\sqrt{3}, 1$ on $[0, 1]$. Substituting these values into the above expression for $f^{(iv)}(t)$ shows that

$$\max_{0 \leqslant t \leqslant 1} \{|f^{(iv)}(t)|\} = f^{(iv)}(0) = 24.$$

Consequently, from (4.31b), an upper bound for the global error in $S(1/16)$ is

$$\frac{1 \cdot (1/16)^4}{180} \cdot 24 \simeq 2 \times 10^{-6},$$

i.e. the global error ε_G associated with the estimate $S(1/16)$ satisfies

$$|\varepsilon_G| \leqslant 0.000002.$$

Notice that this does not confirm that $S(1/16)$ is accurate to six decimal places. In this example, $|\varepsilon_G|$ is considerably smaller than the upper bound obtained from (4.31).

An alternative approach to the previous example would be to use the upper bound for ε_G to determine a step length that would guarantee six-decimal-place accuracy. Thus we must choose h such that the right-hand side of (4.31) is less than 5×10^{-7}, i.e. we require

$$\frac{1 \cdot h^4 \cdot 24}{180} < 5 \times 10^{-7}$$

which leads to

$$h < 0.044.$$

In this example, $nh = 1$ since the interval has length 1 and therefore we require

$$n > \frac{1}{0.044} = 22.7.$$

To apply Simpson's composite formula, n—the number of sub-intervals $[x_i, x_{i+1}]$—

must be even, so that $n + 1$—the number of data points—is odd. The smallest even integer satisfying the above inequality is 24. Hence, applying Simpson's composite formula with step length $h = 1/n = 1/24$, would give six-decimal-place accuracy. This method is usually more efficient than the method used in the above example. However, n is often larger than necessary since its selection is based on an upper bound for the global error.

Algorithm 4.2. Simpson's rule

The interval of integration is $[a, b]$ and the number of intervals is n, which must be even; the function f is to be specified.

read (a, b, n);
$h := (b - a)/n$;
Sum $:= f(a) + f(b) + 4*f(a + h)$;
FOR $i := 1$ TO $n/2 - 1$ DO
      SUM $:= $ Sum $+ 2*f(a + 2*i*h) + 4*f(a + (2*i + 1)*h)$;
IntegVal $:= h*$Sum$/3$;
write (IntegVal).

4.3 TRUNCATION ERROR AND CLASSIFICATION OF INTEGRATION FORMULAE

In the previous section we derived two well-known integration formulae for estimating the definite integral (4.1). There are many more, and consequently it is useful to classify them according to their accuracy. It is normally possible to achieve a required accuracy with any composite formula by taking a sufficiently small step length (as demonstrated in Examples 4.1 and 4.2). However, for a given step length, some formulae are more accurate than others.

To compare integration formulae and examine their accuracy we look at their truncation errors. Local truncation errors are almost always expressed in the form $ch^{r+1} f^{(r)}(\xi)$ where c is a constant, r is a positive integer and $f^{(r)}(\xi)$ is the rth derivative of the integrand $f(x)$ calculated at $x = \xi$. The truncation errors associated with the Trapezium rule and Simpson's rule have already been expressed in this form in equations (4.13) and (4.27) respectively. It can be assumed that x and h are non-dimensional variables in this error term, since the substitution $x = (X - a)/(b - a)$ leads to the equality

$$\int_a^b F(X)\,dX = \int_0^1 f(x)\,dx.$$

Thus if $nH = b - a$ it follows that $nh = 1$ and consequently $h \leqslant 1$ (more briefly, multiply numerator and denominator of $ch^{r+1} f^{(r)}(\xi)$ by $(b - a)^{r+1}$ and redefine x, h, c as $(x - a)/(b - a)$, $h/(b - a)$, $c(b - a)$.

Since $h < 1$ (except for the simplist application of the Trapezium rule with $h = 1$) it is evident that larger values of r lead to smaller errors and more accurate estimates of the integral. An integration formula with truncation error of the form $ch^{r+1}f^{(r)}(\xi)$ is referred to as an rth-order formula, i.e. if the formula was expanded as a power series in h, terms up to and including h^r are included in the formula but terms of order $r+1$ in h and higher are not included. Consequently high-order formulae give more accurate results than low-order formulae.

Composite formulae and their associated global error are derived by adding Newton–Cotes formulae and their associated local truncation errors. It is only to be expected that some accuracy is lost when adding several errors together. We find that the global error is always one order of h lower than the local truncation errors from which it is formed. We have only derived upper bounds for global errors associated with the Trapezium rule (equation (4.18)) and Simpson's rule (equation (4.31)). However, lower bounds can be derived in a similar manner and we find that the global error associated with the composite form of the Trapezium rule satisfies the inequality

$$(b-a)\min_{a\leqslant x\leqslant b}\{|f''(x)|\}\leqslant|\varepsilon_{\text{G}}|\leqslant(b-a)\frac{h^2}{12}\min_{a\leqslant x\leqslant b}\{|f''(x)|\}. \tag{4.32}$$

Consequently the global error for the Trapezium rule has order h^2 whereas the local truncation error has order h^3 (equation (4.13)). Similarly the global error associated with the composite form of Simpson's rule satisfies the inequality

$$(b-a)\frac{h^4}{180}\min_{a\leqslant x\leqslant b}\{|f^{(\text{iv})}(x)|\}\leqslant|\varepsilon_{\text{G}}|\leqslant(b-a)\frac{h^4}{180}\max_{a\leqslant x\leqslant b}\{|f^{(\text{iv})}(x)|\}, \tag{4.33}$$

showing that it has order h^4 whereas the local truncation error for Simpson's rule has order h^5 (equation 4.27).

4.4 HIGHER ORDER INTEGRATION FORMULAE

By taking piecewise-linear and piecewise-quadratic approximations to the integrand $f(x)$ we have derived the Trapezium rule and Simpson's rule for approximating the integral (4.1). The composite forms of these formulae had global errors of order h^2 and h^4. More accurate integration formulae, with error terms of higher order, can be obtained by using higher degree piecewise-polynomial approximations to the integrand. Instead of (4.3) and (4.19) we may use the more general division

$$\int_{x_0}^{x_n}f(x)\,\mathrm{d}x=\sum_{i=0(m)n-m}\int_{x_i}^{x_{i+m}}f(x)\,\mathrm{d}x \tag{4.34}$$

with the approximation

$$\int_{x_i}^{x_{i+m}}f(x)\,\mathrm{d}x=\int_0^m\left\{f_i+r\Delta f_i+\cdots+\frac{r(r-1)(r-m+1)}{m!}\Delta^m f_i\right\}h\,\mathrm{d}r. \tag{4.35}$$

Here a polynomial of degree m is used to approximate the integrand on each sub-interval of the form $[x_i, x_{i+m}]$. The polynomial passes through the data points P_k for $k=i(1)i+m$ and ε_i is the local truncation error. A formula can be obtained for each permissible (see below) value of the positive integer m and a set of formulae can be generated by taking $m=1,2,3,\ldots$ in turn. Equation (4.35) leads to the Newton–Cotes version of each formula and equation (4.34) gives the corresponding composite formula. The first two values of m have already been dealt with and lead to the Trapezium rule and Simpson's rule respectively. The next two Newton–Cotes formulae with their local truncation errors are

$$\int_{x_i}^{x_{i+3}} f(x)\,dx = \frac{3h}{8}\{f_i + 3f_{i+1} + 3f_{i+2} + f_{i+3}\} - \frac{3h^5}{80}f^{(iv)}(\xi_i) \qquad (4.36)$$

and

$$\int_{x_i}^{x_{i+4}} f(x)\,dx = \frac{2h}{45}(7f_i + 32f_{i+1} + 12f_{i+2} + 32f_{i+3} + 7f_{i+4}) - \frac{8h^7}{945}f^{(vi)}(\xi_i). \quad (4.37)$$

These can be obtained from (4.35) by taking $m = 3$ and $m = 4$ in turn and carrying out the elementary integrations. Equation (4.36) is called the three-eighths rule. The corresponding composite formulae can be obtained by substituting (4.36) and (4.37) into (4.34) with $m = 3$ and $m = 4$ respectively.

It was noted in section 4.1 that the division indicated in (4.34) is only possible when n is divisible by m. This is, perhaps, more obvious when (4.34) is expressed in the equivalent form

$$\int_{x_0}^{x_n} f(x)\,dx = \sum_{k=0}^{(n/m)-1} \int_{x_{mk}}^{x_{m(k+1)}} f(x)\,dx, \qquad (4.38)$$

e.g. for the three-eighths rule ($m = 3$) we have

$$\int_{x_0}^{x_n} f(x)\,dx = \sum_{k=0}^{(n/3)-1} \int_{x_{3k}}^{x_{3k+3}} f(x)\,dx \qquad (4.39)$$

and clearly n must be divisible by 3.

4.5 CHOICE OF FORMULA

It is not possible to say that one formula is better than all other formulae for all problems. There are advantages and disadvantages associated with each formula. The condition that n must be divisible by m will rule out many formulae in a given problem. Since $m = 1$ for the Trapezium rule it is only the Trapezium rule that can be used for all problems. However, compared to other formulae, the Trapezium rule has large truncation error, and therefore a small step length is needed. This may involve considerable computation and, in general, it is better to use a higher order method.

The trapezium rule, which corresponds to $m = 1$, is at one extreme among the methods for approximating the integral (4.1). By taking $m = n$ we have the other extreme and this yields the highest order method. It is the Newton–Cotes formula applied to the whole interval $[x_0, x_n]$ and results from approximating the integrand on $[x_0, x_n]$ by the polynomial of degree n that passes through the $n + 1$ data points. If n is large, the integration formula will be extremely complicated. For example, the Trapezium rule approximation to the integral over $[x_0, x_8]$ is

$$\int_{x_0}^{x_8} f(x)\,dx \simeq h\left\{\frac{f_0}{2} + f_1 + f_2 + f_3 + f_4 + f_5 + f_6 + f_7 + \frac{f_8}{2}\right\} \qquad (4.40)$$

and the Newton–Cotes approximation is

$$\int_{x_0}^{x_8} f(x)\,dx \simeq \frac{4h}{14175}(989 f_0 + 5888 f_1 - 928 f_2 + 10496 f_3$$

$$- 4540 f_4 + 10496 f_5 - 928 f_6 + 5888 f_7 + 989 f_8). \qquad (4.41)$$

However the Trapezium rule has associated truncation error of order h^2 whereas (4.41) has truncation error of order h^{11}. It is rarely necessary to use formulae such as (4.41) which are computationally inefficient. Clearly a compromise between the two extremes $m = 1$ and $m = n$ is preferable.

Simpson's rule is a good compromise. In composite form its associated truncation error is of order h^4 compared to h^2 for the composite form of the Trapezium rule. Indeed it is a bonus that Simpson's rule is of order h^2 better than the Trapezium rule since they were derived from piecewise-quadratic and piecewise-linear approximations to the integrand respectively. There is no advantage in using the three-eighths rule (4.36) rather than Simpson's rule, since the truncation errors are of the same order. However the three-eighths rule can be used in conjunction with Simpson's formula to remove the restriction that n must be even. Thus if n is even (there are $n + 1$ data points), use Simpson's composite formula, and if n is odd, use a combination of the three-eighths rule and Simpson's rule. For example, the range $[x_0, x_n]$ could be split into $[x_0, x_3]$ and $[x_3, x_n]$. Then, the three-eighths rule can be applied to $[x_0, x_3]$ and, since $n - 3$ is odd, Simpson's composite formula can be applied to $[x_3, x_n]$.

When combining different integration formulae, careful consideration must be given to the truncation errors—global or local, as appropriate—to ensure that accuracy is not lost. For example, when a formula is being used locally to complement Simpson's composite formula, its truncation error should be of order h^4 or higher. Above, we suggested use of the three-eighths rule locally to complement Simpson's composite formula when n is odd. The local truncation error associated with the three-eighths rule is of order h^5, which is better than required. An alternative to the three-eighths rule that overcomes the problem of n being odd is to use the Trapezium rule locally. However this is an unsuitable alternative because the local truncation error for the Trapezium rule is only of order h^3, and overall accuracy may be lost.

The above comments are fairly general. Specific problems have their own peculiarities which may influence the choice of integration formula. In the next section we meet a method that can yield high-order results but is based on a simple formula.

4.6 ROMBERG INTEGRATION

This is a method of obtaining high-order numerical approximations to integrals from Trapezium rule (or Simpson's rule) approximations. To obtain a higher order method based on the Trapezium rule we examine the correction or error term associated with the Trapezium rule. We first show that the correction term associated with the composite form of the Trapezium rule can be expressed in the form

$$C = a_2 h^2 + a_4 h^4 + a_6 h^6 + \cdots, \qquad (4.42)$$

where h is the step length and $a_2, a_4, a_6, \ldots$ are constants that are of no interest to us in this particular application.

To justify the expansion (4.42) we show that the correction term is an even function of h. Since the correction to the Trapezium rule is simply the difference between the exact integral and the Trapezium rule approximation we have

$$C = \int_{x_0}^{x_n} f(x)\,dx - h\left\{\frac{f_0}{2} + f_1 + f_2 + \cdots + f_{n-1} + \frac{f_n}{2}\right\}. \tag{4.43}$$

Considering the two terms on the right of (4.43) separately we first note that

$$\left\{\frac{f_0}{2} + f_1 + \cdots + f_{n-1} + \frac{f_n}{2}\right\} = (1 + E + E^2 + \cdots + E^{n-1})f_0 - \frac{f_0}{2} + \frac{f_n}{2}$$

$$= \left(\frac{E^n - 1}{E - 1}\right)f_0 + \frac{(E^n - 1)}{2}f_0,$$

where E is the shift operator (see equations (3.7)–(3.9)). Also, if $F(x)$ is the indefinite integral of $f(x)$ we may write

$$\int_{x_0}^{x_n} f(x)\,dx = F(x_n) - F(x_0)$$

$$= [(E^n - 1)F(x)]_{x = x_0}$$

$$= \left[(E^n - 1)\frac{1}{D}f(x)\right]_{x = x_0}$$

where D is the differential operator. Hence, using the operator identity $E = e^{hD}$ (see Exercise 3), equation (4.43) can be written

$$C(h) = \left[(E^n - 1)\left\{\frac{1}{D} - \frac{h}{e^{hD} - 1} - \frac{h}{2}\right\}f(x)\right]. \tag{4.44}$$

It remains to show that $C(-h) = C(h)$. We have

$$C(-h) = \left[(E^n - 1)\left\{\frac{1}{D} + \frac{h}{e^{-hD} - 1} + \frac{h}{2}\right\}f(x)\right]_{x = x_0}$$

and

$$\frac{1}{D} + \frac{h}{e^{-hD} - 1} + \frac{h}{2} = \frac{1}{D} + \frac{he^{hD}}{1 - e^{hD}} + \frac{h}{2}$$

$$= \frac{1}{D} + \left\{\frac{he^{hD}}{1 - e^{hD}} + h\right\} - h + \frac{h}{2}$$

$$= \frac{1}{D} - \frac{h}{e^{hD} - 1} - \frac{h}{2},$$

giving

$$C(-h) = \left[(E^n - 1)\left\{\frac{1}{D} - \frac{h}{e^{hD} - 1} - \frac{h}{2}\right\}f(x)\right]_{x = x_0}$$

which is the same as (4.44), i.e. $C(-h) = C(h)$ as required and equation (4.42) is justified.

Denoting the exact integral by I and the Trapezium rule approximation with step length h by $T(h)$, from (4.42) and (4.43) we may write

$$I = T(h) + a_2h^2 + a_4h^4 + a_6h^6 + \cdots. \tag{4.45}$$

If (4.45) is used with step lengths h and $h/2$ we have

$$I = T(h) + a_2h^2 + a_4h^4 + \cdots$$

and

$$I = T(h/2) + a_2\frac{h^2}{4} + a_4\frac{h^4}{16} + \cdots.$$

Eliminating the leading contribution to the correction term between these two equations gives

$$I = \frac{4T(h/2) - T(h)}{3} + b_4h^4 + \cdots, \tag{4.46}$$

where b_4 is a constant whose value is of no interest to us. However we see that, whereas $T(h)$ and $T(h/2)$ have errors of order h^2, the first term on the right-hand side of (4.46),

$$T(h, h/2) = \frac{2^2T(h/2) - T(h)}{2^2 - 1}, \tag{4.47}$$

gives an approximation to I with error of order h^4. Clearly, little effort is required to evaluate this higher order approximation once $T(h)$ and $T(h/2)$ are known.

When applying Romberg integration the coefficients involved are most easily recalled when they are expressed in the form given in (4.47), i.e. $2^2, -1, 2^2 - 1$. This will become more evident as we derive even higher order approximations. In general we will find that the coefficients are 2^{2n} and -1 in the numerator and $2^{2n} - 1$ in the denominator when we have an approximation to I of order $2n + 2$, where n is a positive integer. Notice that the right-hand side of (4.47) gives more weight to the Trapezium rule approximation with the smaller step length. A similar feature arises in the higher order approximations (see below).

From (4.46) and (4.47) we have

$$I = T(h, h/2) + b_4h^4 + b_6h^6 + \cdots$$

and replacing h by $h/2$ gives

$$I = T(h/2, h/4) + b_4\frac{h^4}{16} + b_6\frac{h^6}{64} + \cdots,$$

where, replacing h by $h/2$ in (4.47),

$$T(h/2, h/4) = \frac{2^2T(h/4) - T(h/2)}{2^2 - 1}. \tag{4.48}$$

Eliminating the leading contribution to the error term between these two expansions for I gives

$$I = \frac{16T(h/2, h/4) - T(h, h/2)}{15} + c_6h^6 + \cdots. \tag{4.49}$$

Hence

$$T(h, h/2, h/4) = \frac{2^4T(h/2, h/4) - T(h, h/2)}{2^4 - 1} \tag{4.50}$$

gives an approximation to I with error of order h^6. Observe that the three Trapezium rule approximations $T(h)$, $T(h/2)$ and $T(h/4)$ must be calculated in order that (4.50) can be evaluated (also see (4.47), (4.48)).

This procedure can be extended indefinitely. Thus, if $T(h)$, $T(h/2)$, $T(h/4)$ and $T(h/8)$ are calculated, we can also calculate

$$T(h/2, h/4, h/8) = \frac{2^4 T(h/4, h/8) - T(h/2, h/4)}{2^4 - 1}, \tag{4.51}$$

which is (4.50) with h replaced by $h/2$. Then it can be shown that

$$T(h, h/2, h/4, h/8) = \frac{2^6 T(h/2, h/4, h/8) - T(h, h/2, h/4)}{2^6 - 1} \tag{4.52}$$

gives an approximation to I with error of order h^8.

When using Romberg integration it is convenient to construct a table which takes the form

$$
\begin{array}{cccc}
h^2 & h^4 & h^6 & h^8 \\
T(h) & & & \\
& T(h, h/2) & & \\
T(h/2) & & T(h, h/2, h/4) & \\
& T(h/2, h/4) & & T(h, h/2, h/4, h/8) \\
T(h/4) & & T(h/2, h/4, h/8) & \\
& T(h/4, h/8) & & \\
T(h/8) & & &
\end{array}
$$

The heading in each column gives the order of the error associated with the entries in that column. To summarize, the first column is calculated directly from the Trapezium rule using the step lengths indicated. The remaining columns can be easily calculated via equations such as (4.47), (4.50), (4.52), each entry depending on the two closest values in the column immediately to the left.

Example 4.3 Use the Trapezium rule and Romberg integration to estimate

$$\int_0^{0.8} e^{x^2} \, dx.$$

Here we are repeating Example 4.1. We demonstrate that the same accuracy can be achieved with less calculation. In Example 4.1 we found that it was necessary to calculate $I(0.4) = 1.049$, $I(0.2) = 1.019$, $I(0.1) = 1.012$ and $I(0.5) = 1.010$ to obtain two-decimal-place accuracy. However using Romberg integration we find that only the first three of these values are needed to obtain the same accuracy. The Romberg table takes the form

$$
\begin{array}{cc}
h^2 & h^4 \\
I(0.4) = 1.049 & \\
& 1.009 \\
I(0.2) = 1.019 & \\
& 1.010 \\
I(0.1) = 1.012 &
\end{array}
$$

The second column gives two-decimal-place accuracy.

Alternatively, if we make use of all four values in Example 4.1 and apply Romberg integration we obtain a more accurate result with little additional calculation. The Romberg table takes the form

h^2	h^4	h^6
$I(0.4) = 1.0487005$		
	1.0093376	
$I(0.2) = 1.0191783$		1.0091215
	1.0091350	
$I(0.1) = 1.0116458$		1.0091207
	1.0091216	
$I(0.05) = 1.0097526$		

This yields the estimate 1.00912 to five decimal places, which is a considerable improvement over the result obtained in Example 4.1.

Algorithm 4.3. The Romberg table

The interval of integration is $[a, b]$ and the initial number of intervals is n; the function f is to be specified. The tolerance is Tol.

```
read(a, b, n, Tol);
h:= (b − a)/n;
Sum:= (f(a) + f(b))/2
FOR i:= 1 TO n − 1 DO Sum + f(a + i*h);
Trap[1, 1]:= h*Sum;
J:= 1;REPEAT
   FOR k:= O TO n − 1 DO Sum:= Sum + f(a + k*h + h/2);
   n:= 2*n; j:= j + 1; Factor:= 1; h:= h/2;
   Trap[1, j]:= h*Sum;
   FOR i:= 1 TO j − 1 DO
     BEGIN
     Factor:= 4*Factor;
     Trap[i + 1, j]:= (Factor*Trap[i, j] − Trap[i, j − 1])/(Factor − 1)
     END
   UNTIL abs (Trap[j,j] − Trap[j − 1, j]) < Tol;
FOR k:= 1 TO j DO
   FOR:= 1 TO k DO write(Trap[i, k]).
```

4.7 CLOSED AND OPEN INTEGRATION FORMULAE

In our derivations of Newton–Cotes integration formulae for estimating

$$\int_{x_0}^{x_n} f(x)\,dx$$

we have approximated the integrand $f(x)$ by the polynomial that passes through the data points (x_i, f_i) for $i = 0(1)n$. In particular the polynomial approximation to $f(x)$ passes through the end values (x_0, f_0) and (x_n, f_n) as well as the interior data points (x_i, f_i) for $i = 1(1)n - 1$. Consequently an integration formula is obtained in which the end values, f_0 and f_n, appear explicitly. These formulae, that depend on the values of the integrand at the ends of the integration range, are called closed integration formulae.

Some problems arise in which the values of the integrand are not known at the ends of the range of integration. In this situation the above integral must be estimated by a formula that does not depend on f_0 and f_n. A formula of this type can be derived by approximating the integrand $f(x)$ over the range $[x_0, x_n]$ by the polynomial passing through the interior points (x_i, f_i) for $i = 1(1)n - 1$. Since the approximation to the integrand does not pass through (x_0, f_0) and (x_n, f_n) the resulting integration formula will not depend on f_0 or f_n. Such formulae are called open integration formulae.

One of the simplest open integration formulae can be obtained by considering the case $n = 3$ for the above integral. With $n = 3$ there are four data points $(x_0, f_0), (x_1, f_1),$ (x_2, f_2) and (x_3, f_3) over the range of integration, $[x_0, x_3]$. To obtain an open integration formula we approximate the integrand over the entire range $[x_0, x_3]$ by the polynomial that passes through the interior points (x_1, f_1) and (x_2, f_2). This is a straight line and can be expressed in the form $f_1 + r\Delta f_1$ where r is defined by $x = x_1 + rh$ (see section 3.2). Thus we make the approximation

$$\int_{x_0}^{x_3} f(x)\,dx \simeq \int_{-1}^{2} \{f_1 + r\Delta f_1\}\,h\,dr. \tag{4.53}$$

Note that the limits $r = -1$ and $r = 2$ correspond to the limits $x = x_0$ and $x = x_3$, and that this correspondence arises from the relationship $x = x_1 + rh$. Carrying out the elementary integrations and substituting $\Delta f_1 = f_2 - f_1$ gives

$$\int_{x_0}^{x_3} f(x)\,dx \simeq \frac{3h}{2}(f_1 + f_2).$$

This formula is not sufficiently accurate for most applications. It is only presented as a simple illustration of an open formula.

Another open formula is Milne's formula:

$$\int_{x_0}^{x_4} f(x)\,dx = \frac{4h}{3}(2f_1 - f_2 + 2f_3) + \tfrac{14}{45}h^5 f^{(iv)}(\xi). \tag{4.54}$$

Here the integration interval is $[x_0, x_4]$ and there are five points (x_i, f_i), $i = 0(1)4$, over this range. The formula is derived by approximating the integrand over $[x_0, x_4]$ by the quadratic passing through the three interior points (x_i, f_i), $i = 1(1)3$. As with closed integration formulae the error term can be derived by examining the term omitted from the interpolation formula that is used to approximate the integrand.

Milne's formula has been used in conjunction with Simpson's formula, which is closed, to solve ordinary differential equations. However, other formulae, with better stability properties, are normally preferred.

Our main reason for describing open formulae is because of their use in the numerical solution of ordinary differential equations. However, in this application, the integration formulae are often closed at one end of an interval, where the solution is known, and open at the other end, where the solution is not known. To obtain such a formula to estimate the integral in (4.53) the integrand could be approximated over $[x_0, x_4]$ by the polynomial passing through (x_i, f_i), $i = 0(1)3$. The resulting formula would contain f_0, f_1, f_2 and f_3 but not f_4. This is discussed in Chapter 5.

EXERCISES 4

1. Use the Trapezium rule with four step lengths to estimate the following integrals. Compare your estimates with the exact results.

   (a) $\displaystyle\int_0^2 x^2\,dx$ (b) $\displaystyle\int_1^2 \frac{1}{x}\,dx$

   (c) $\displaystyle\int_0^1 \cos\frac{\pi x}{2}\,dx$ (d) $\displaystyle\int_0^2 \frac{x}{(2+x^2)^{1/2}}\,dx.$

2. Repeat Exercise 1 using the composite form of Simpson's rule and four step lengths.

3. By repeated application of the Trapezium rule with successively smaller step lengths estimate

$$\int_{0.5}^2 e^{-x^2}\,dx$$

   to three decimal places.

4. Estimate $\cosh^{-1}$ to four decimal places by repeated application of the Trapezium rule to the derivative of $\coth^{-1} x$.

5. Use Simpson's rule to estimate

$$\int_0^1 \frac{\sin x}{2+3\sin x}\,dx$$

   with a step length of 0.25 initially. Repeat the calculation as necessary to estimate the integral to three decimal places.

6. Estimate

$$\int_0^1 \frac{\sin x}{x}\,dx$$

   to four significant figures using

   (i) the Trapezium rule
   (ii) Simpson's rule.

7. By using the upper bound for the global truncation error in the Trapezium rule, determine the value of the step length that will guarantee four-decimal-place accuracy when applying the composite Trapezium rule to the integrals in Exercise 1.

8. Repeat Exercise 7 using Simpson's rule instead of the Trapezium rule.

9. If the function $f(x)$ is approximated by a piecewise-cubic polynomial passing through the equally spaced data (x_i, f_i), $i = 0(1)n$, derive the Newton–Cotes and composite formulae for approximating

$$\int_{x_i}^{x_{i+3}} f(x)\,dx \quad \text{and} \quad \int_{x_0}^{x_n} f(x)\,dx.$$

10. Use the Gregory–Newton backward interpolation polynomial (equation (3.31)) and Stirling's interpolation polynomial (equation (3.57b)) to obtain two forms of the quadratic passing through (x_i, f_i), (x_{i+1}, f_{i+1}) and (x_{i+2}, f_{i+2}). Derive Simpson's rule from each of these forms of the quadratic.

11. Derive Milne's open formula, with error term,

$$\int_{x_{i-1}}^{x_{i+3}} f(x)\,dx = \frac{4h}{3}(2f_i - f_{i+1} + 2f_{i+2}) + \tfrac{14}{15}h^5 f^{(iv)}(\xi),$$

where h is the spacing between successive x-values.

12. Derive an integration formula for estimating

$$\int_{x_0}^{x_4} f(x)\,dx$$

when the integrand is only known at the points (x_i, f_i) for $i = 0(1)3$.

13. Show that the difference between $\int_{x_0}^{x_n} f(x)\,dx$ and the composite Trapezium rule can be expressed in the form

$$C = -\frac{h}{12}(\Delta f_n - \Delta f_0) + \frac{h}{24}(\Delta^2 f_n - \Delta^2 f_0) - \tfrac{19}{720}h(\Delta^3 f_n - \Delta^3 f_0) + \cdots.$$

Comment on the use of this formula.

14. By first showing that the forward and backward difference operators are related by

$$\Delta = \nabla + \nabla^2 + \nabla^3 + \cdots,$$

Showing that the correction term in Exercise 9 can be expressed in the more useful form

$$C = \frac{h}{12}(\Delta f_0 - \nabla f_n) - \frac{h}{24}(\Delta^2 f_0 - \nabla^2 f_n) + \frac{19}{720}h(\Delta^3 f_0 - \nabla^3 f_n) - \cdots.$$

This is called the Gregory–Newton correction term.

15. A function $f(x)$ takes the values 0.4401, 0.4709, 0.4983, 0.5220, 0.5419, 0.5579 for $x = 1.0(0.1)1.5$. Estimate

$$\int_1^{1.5} f(x)\,dx$$

using the Trapezium rule with Gregory–Newton correction terms and state the accuracy of your answer.

16. Bessel functions of the first kind, $J_v(x)$, satisfy the equation

$$x^v J_v(x) = \int_0^x t^v J_{v-1}(t) \, dt.$$

For $x = 0.0(0.2)1.0$, $J_{3/2}(x)$ takes the values 0.00000, 0.02369, 0.06621, 0.11921, 0.17840, 0.24030. Using the Trapezium rule with Gregory–Newton correction term, estimate a value for $J_{5/2}(1)$, stating the accuracy of your answer.

17. To obtain a numerical approximation to the integral

$$\int_{-\infty}^{\infty} e^{-x^2} \, dx$$

the infinite limits must be replaced by finite numbers. Determine the integer limits that must be used to obtain four-decimal-place accuracy. Use the Trapezium rule with Romberg integration to estimate this integral to four decimal places. (Its exact value is $\sqrt{\pi}$.)

18. Assuming that

$$I = S(h) + a_4 h^4 + a_6 h^6 + \cdots$$

where

$$I = \int_a^b f(x) \, dx$$

and $S(h)$ is Simpson's rule approximation to I, derive the Romberg approximation

$$S(1, 1/2) = \frac{2^4 S(h/2) - S(h)}{2^4 - 1},$$

which has error of order h^6. Speculate the form the Romberg approximation $S(1, 1/2, 1/4)$, which has error of order h^8.

19. Use Simpson's rule with Romberg integration to estimate

$$\int_0^1 \frac{1}{1 + t^2} \, dt$$

to four decimal places.

20. Repeat Exercise 17 using Simpson's rule with Romberg integration and estimating the integral to six decimal places.

5

Numerical Solution of Ordinary Differential Equations

The solution of differential equations is crucial throughout engineering and science. Many problems are approached by constructing mathematical models. The physical and chemical processes influencing a given problem are considered and incorporated into a set of 'balance' equations. These equations usually involve the rates of change of the variables relating to the problem, i.e. the mathematical model is a set of differential equations and their solution gives an approximate description of the real system. Thus it is necessary to be able to solve these differential equations to investigate the behaviour of the real system.

Some differential equations can be solved analytically. In many problems this only becomes possible by making approximations when constructing the mathematical model. Consequently the solution may not give an accurate description of the system. A better description can be obtained by constructing a more detailed mathematical model. This usually leads to differential equations that cannot be solved analytically. In many cases the differential equations are non-linear. Consequently it is frequently necessary to use numerical techniques to solve them.

In this chapter we consider the numerical solution of ordinary differential equations. Initially we concentrate on a single first-order equation, which can be expressed in the form

$$\frac{dy}{dx} = f(x, y), \tag{5.1}$$

where $y = y(x)$ is a function of the independent variable x and $f = f(x, y)$ is a function of x and y. From an initial condition, which can be expressed in the form

$$y(x_0) = y_0, \tag{5.2}$$

we develop methods for estimating $y(x_1) = y(x_0 + h)$, $y(x_2) = y(x_1 + h), \ldots$, in turn, where h is a suitably chosen step length. Later in the chapter these methods are extended to systems of first-order equations together with a set of initial conditions. A system of first-order equations is usually expressed in the form

$$\frac{dy_1}{dx} = f_1(x, y_1, y_2, \ldots, y_n)$$

$$\frac{dy_2}{dx} = f_2(x, y_1, y_2, \ldots, y_n) \tag{5.3}$$

$$\vdots$$

$$\frac{dy_n}{dx} = f_n(x, y_1, y_2, \ldots, y_n)$$

with the initial conditions

$$y_1(x_0) = \alpha_1, y_2(x_0) = \alpha_2, \ldots, y_n(x_0) = a_n. \tag{5.4}$$

We show that the same techniques can be applied to higher order equations together with a set of initial conditions. This is because the nth-order differential equation

$$\frac{d^n y}{dx^n} = f\left(x, y, \frac{dy}{dx}, \ldots, \frac{d^{n-1} y}{dx^{n-1}}\right) \tag{5.5}$$

can be reduced to a set of first-order equations of the form (5.3). The initial conditions for the nth-order equation usually take the form

$$y(a) = \alpha_0, y^{(1)}(a) = \alpha_1, \ldots, y^{(n-1)}(a) = \alpha_{n-1}, \tag{5.6}$$

where $y^{(k)}(a)$ is the kth derivative of y evaluated at $x = a$. Note that both a and x_0 are commonly used to denote the beginning of the x-interval over which the integration is to take place.

The system of differential equations (5.3) together with the initial conditions (5.4) and the nth-order equation (5.5) together with the initial conditions (5.6) are termed initial-value problems. Thus in an initial-value problem all of the conditions that supplement the differential equations are specified at the beginning of the x-interval. However, problems do arise in which some of the conditions occur at $x = a$ and the others occur at $x = b$, where $[a, b]$ is the interval over which the solution is required. Such problems are termed boundary-value problems. The techniques for solving boundary-value problems are not described in this text. We only examine initial-value problems.

Returning to the first-order equation (5.1) together with the initial condition (5.2) we will assume that we have obtained estimates for $y_1 = y(x_1)$, $y_2 = y(x_2), \ldots, y_n = y(x_n)$. We develop methods that lead to an estimate for $y_{n+1} = y(x_{n+1})$ from these data. These methods can be divided into two categories: multi-step methods and single-step methods. In multi-step methods the formula for estimating y_{n+1} depends not only on the estimate for $y(x_n)$ but also on estimates of y at some of $x = x_{n-1}, x = x_{n-2}, \ldots$. We will see that multi-step methods are normally predictor-corrector-type methods. These methods consist of a predictor formula which gives an initial estimate of the solution and a corrector formula which improves on this initial estimate. With single-step methods the calculation of the estimate for $y(x_{n+1})$ depends only on the previous estimate for $y(x_n)$ and some intermediate quantities calculated over the interval $[x_n, x_{n+1}]$.

5.1 MULTI-STEP METHODS

To develop formulae that give the approximate numerical solution of a differential equation we first rewrite the differential equation as an integral equation. This transforms the problem to one of approximating an integral, and techniques similar to those described in Chapter 4 can be used. Thus, assuming that estimates $w_1, w_2, \ldots, w_n$ for $y_1, y_2, \ldots, y_n$ have been calculated, we integrate equation (5.1) over the interval $[x_n, x_{n+1}]$ to obtain

$$y(x_{n+1}) = y(x_n) + \int_{x_n}^{x_{n+1}} f(x, y) \, dx. \qquad (5.7)$$

To develop integration formulae from (5.7) we need to approximate the integrand $f(x, y)$ over the interval $[x_n, x_{n+1}]$. At this stage a value for $f(x, y)$ at $x = x_{n+1}$ is not available since the estimate $w_{n+1} \simeq y_{n+1}$ is not known. Thus we need an integration formula that does not contain $f_{n+1} = f(x_{n+1}, y_{n+1}) \simeq f(x_{n+1}, w_{n+1})$, i.e. an integration formula that is open with respect to $x = x_{n+1}$ is required (see section 4.7). This is achieved by approximating the integrand by a polynomial that does not pass through the point (x_{n+1}, f_{n+1}), i.e. the integrand must be approximated by a polynomial passing through some of the previously estimated points (x_i, f_i), $i = 0(1)n$. Using an approximation of this type in (5.7) leads to an estimate w_{n+1} for $y(x_{n+1})$. One this is known, $f_{n+1} = f(x_{n+1}, y_{n+1})$ can also be estimated and an alternative approximation to the integrand $f(x, y)$ can be obtained using a polynomial that passes through (x_{n+1}, f_{n+1}) as well as some of (x_i, f_i), $i = 0(1)n$. This leads to a closed integration formula and a new estimate for $y(x_{n+1})$ via equation (5.7). This new estimate is normally more accurate than the initial estimate.

The closed-integration approach cannot be used without an initial estimate for f_{n+1}. Thus, in practice, formulae that result from open integration are used to supply an initial estimate, which is then improved by formulae that result from closed integration. The formulae used to obtain the initial estimate are called predictor formulae and the formulae that improve the initial estimates are called corrector formulae. This leads to the so-called predictor–corrector methods.

5.1.1 Methods based on open integration

Since estimates of $y_1, y_2, \ldots, y_n$ are already available, estimates of the integrated $f_i = f(x_i, y_i)$ can also be calculated for $i = 0(1)n$. Consequently the integrand $f(x, y)$ can be approximated by the Nth-degree polynomial passing through the last $N + 1$ points, (x_i, f_i) for $i = n(-)n - N$ $(N < n)$. In these circumstances the most appropriate form of this polynomial is the Gregory–Newton backward interpolation polynomial based at $x = x_n$ (see equation (3.31)). Approximating the integrand in this way, equation (5.7) takes the form

$$y(x_{n+1}) = y(x_n) + \int_0^1 \left\{ f_n + r\nabla f_n + \frac{r(r+1)}{2!} \nabla^2 f_n + \cdots \right.$$

$$\left. + \frac{r(r+1)\cdots(r+N-1)}{N!} \nabla^N f_n \right\} h \, dr + \varepsilon_{n+1} \qquad (5.8)$$

where $x = x_n + rh$ and ε_{n+1} is the error associated with the approximation of the integrand by this polynomial, i.e. ε_{n+1} is the error resulting from the termination of the Gregory–Newton backward formula after the term in the Nth order difference and is given by

$$\varepsilon_{n+1} = \int_0^1 \left\{ \frac{r(r+1)\cdots(r+N)}{(N+1)!} \nabla^{N+1} f_n \cdots \right\} h \, dr. \tag{5.9}$$

If ε_{n+1} is omitted from (5.8) the solution of the resulting difference equation will be an approximation to $y(x_n)$. Denoting this approximation by w_n equation (5.8) can be written

$$w_{n+1} = w_n + h \sum_{k=0}^{N} a_k \nabla^k f(x_n, w_n) \tag{5.10}$$

where

$$a_0 = 1, \quad a_k = \int_0^1 \frac{r(r+1)\cdots(r+k-1)}{k!} dr \quad \text{for } k = 1(1)N. \tag{5.11}$$

Evaluating the elementary integral (5.11) for the first few values of k shows that (5.10) can be written

$$\begin{aligned} w_{n+1} = w_n + h\{ f_n + \tfrac{1}{2}\nabla f_n + \tfrac{5}{12}\nabla^2 f_n + \tfrac{3}{8}\nabla^3 f_n \\ + \tfrac{251}{720}\nabla^4 f_n + \tfrac{95}{288}\nabla^5 f_n + \cdots + a_N \nabla^N f_n \}. \end{aligned} \tag{5.12}$$

Taking $N = 0, 1, 2, 3, 4$ in turn and using equation (3.24) to express the backward differences in terms of function values (also see Example 3.2) leads to the following integration schemes:

First-order, one-step

$$w_{n+1} = w_n + h f_n, \quad \varepsilon_{n+1} = \frac{h^2}{2} y^{(2)}(\xi_n). \tag{5.13}$$

Second-order, two-step

$$w_{n+1} = w_n + \frac{h}{2}(3f_n - f_{n-1}), \quad \varepsilon_{n+1} = \tfrac{5}{12}h^3 y^{(3)}(\xi_n). \tag{5.14}$$

Third-order, three-step

$$w_{n+1} = w_n + \frac{h}{12}(23f_n - 16f_{n-1} + 5f_{n-2}),$$

$$\varepsilon_{n+1} = \tfrac{3}{8}h^4 y^{(4)}(\xi_n). \tag{5.15}$$

Fourth-order, four-step

$$w_{n+1} = w_n + \frac{h}{24}(55f_n - 59f_{n-1} + 37f_{n-2} - 9f_{n-3}),$$

$$\varepsilon_{n+1} = \tfrac{251}{720}h^5 y^{(5)}(\xi_n). \tag{5.16}$$

Fifth-order, five-step

$$w_{n+1} = w_n + \frac{h}{720}[1901 f_n - 2774 f_{n-1} + 2616 f_{n-2} - 1274 f_{n-3} + 251 f_{n-4}]$$

$$\varepsilon_{n+1} = \tfrac{95}{288} h^6 y^{(6)}(\xi_n). \tag{5.17}$$

The order of these schemes is defined in the same way as for integration formulae in Chapter 4. A scheme with error term containing the factor h^{k+1} has order k. Clearly higher order schemes can be obtained by taking larger values of N in (5.12).

All of these formulae that result from equation (5.12) are called Adams–Bashforth formulae. They were derived by open integration and a consequence of this is that they lead to an explicit expression for $w_{n+1} \simeq y(x_{n+1})$ in terms of previous estimates of y and $f(x, y)$. Thus in a given problem, when $f(x, y)$ is known, these formulae are easy to use to advance the solution from $x = x_n$ to $x = x_{n+1}$.

The Adams–Bashforth formulae have been derived by truncating equation (5.12). The term after which (5.12) was truncated was determined by the value of N. In each case the form of the error, associated with this truncation, is given. Although not rigorous, the correct form of these error terms can be obtained in the same way as error terms were obtained for numerical integration formulae in Chapter 4, i.e. by examining the leading contribution to the terms omitted from (5.12). This is evident when comparing the coefficients in each error term of equations (5.13)–(5.17) with equation (5.12). As an example consider the derivation of equation (5.15). This is obtained by taking $N = 2$ in (5.12), giving

$$w_{n+1} = w_n + h\{f_n + \tfrac{1}{2}\nabla f_n + \tfrac{5}{12}\nabla^2 f_n\}, \tag{5.18}$$

which leads to (5.15) on substituting $\nabla f_n = f_n - f_{n-1}$ and $\nabla^2 f_n = f_n - 2f_{n-1} + f_{n-2}$. From (5.9) the error associated with this formula is

$$\varepsilon_{n+1} = \int_0^1 \left\{ \frac{r(r+1)(r+2)}{3!} \nabla^3 f_n + \cdots \right\} h\, dr. \tag{5.19}$$

The coefficient of $\nabla^3 f_n$ here is (5.11) with $k = 3$ and this has been evaluated in (5.12), i.e.

$$\varepsilon_{n+1} = h\{\tfrac{3}{8}\nabla^3 f_n + \cdots\}. \tag{5.20}$$

To express the error in terms of derivatives we use the relations

$$\nabla = hD - \frac{h^2 D^2}{2!} + \frac{h^3 D^3}{3!} - \cdots \tag{5.21}$$

and

$$\nabla^s = h^s D^s + \cdots. \tag{5.22}$$

The reader is left to derive these relations between the backward difference operator ∇ and the differential operator D (Exercise 4). Similar results have been derived involving the forward difference operator Δ in equations (4.10), (4.11). Using (5.22) with $s = 3$, (5.20) can be written

$$\varepsilon_{n+1} = \tfrac{3}{8} h^4 f^{(3)}(x_n) + \cdots. \tag{5.23}$$

A more rigorous approach shows that this leading term gives the precise form of the error when the derivative is evaluated at a different value of x, ξ_n say, i.e. it can be shown that

$$\varepsilon_{n+1} = \tfrac{3}{8}h^4 f^{(3)}(\xi_n). \tag{5.24}$$

Further, since $dy/dx = f(x, y)$ (equation (5.1)), equation (5.24) can also be written as

$$\varepsilon_{n+1} = \tfrac{3}{8}h^4 y^4(\xi_n). \tag{5.25}$$

The error terms for the other schemes can be obtained similarly.

By examining the error terms associated with the schemes (5.13)–(5.17) it is evident that, in general, larger values of N lead to more accurate formulae. This is because the approximation to the integrand is the source of the error, and larger values of N given better approximations to the integrand. In equation (5.8) the integrand has been approximated by the Nth-degree polynomial passing through $(x_n, f_n), (x_{n-1}, f_{n-1}), \dots,$ (x_{n-N}, f_{n-N}). Thus as N increases, the polynomial approximation passes through more of the previous data points, leading to smaller errors, as expected.

5.1.2 Methods based on closed integration

Once a value for $y(x_{n+1})$ has been estimated by any of the schemes derived in section 5.1.1, $f_{n+1} = f(x_{n+1}, y_{n+1})$ can be estimated. The integrand $f(x, y)$ in (5.7) can then be approximated over $[x_n, x_{n+1}]$ by a polynomial passing through (x_{n+1}, f_{n+1}) as well as some of the previous points (x_k, f_k) for $k = 0(1)n$. This will lead to a closed-integration formula for the approximation of the integral.

The most convenient form of the Nth-degree polynomial passing through the last $N + 1$ data points (x_k, f_k) for $k = n + 1(-1)n - N + 1$ is the Gregory–Newton backward formula based at $x = x_{n+1}$ (equation (3.31) with $j = n + 1$). Using the polynomial to approximate the integrated, equation (5.7) takes the form

$$y(x_{n+1}) = y(x_n) + \int_{-1}^{0} \left\{ f_{n+1} + s\nabla f_{n+1} + \frac{s(s+1)}{2!}\nabla^2 f_{n+1} \right.$$
$$\left. + \dots + \frac{s(s+1)\cdots(s+N-1)}{N!}\nabla^N f_{n+1} \right\} h\,ds + \varepsilon_{n+1}, \tag{5.26}$$

where s is defined by $x = x_{n+1} + sh$ and ε_{n+1} is the error associated with the approximation of the integrand by this polynomial. Thus ε_{n+1} is the error resulting from the termination of the Gregory–Newton backward formula after the term in the Nth-order difference and is given by

$$\varepsilon_{n+1} = \int_{-1}^{0} \left\{ \frac{s(s+1)\cdots(s+N)}{(N+1)!}\nabla^{N+1} f_{n+1} \cdots \right\} h\,ds. \tag{5.27}$$

If ε_{n+1} is omitted from (5.26) the solution of the resulting difference equation will be an approximation to $y(x_n)$. Denoting this approximation by w_n equation (5.26) can be written

$$w_{n+1} = w_n + h \sum_{k=0}^{N} b_k \nabla^k f(x_{n+1}, w_{n+1}), \tag{5.28}$$

where

$$b_0 = 1, \quad b_k = \int_{-1}^{0} \frac{s(s+1)\cdots(s+N-1)}{N!}\,ds \quad \text{for } k = 1(1)N. \tag{5.29}$$

Carrying out the elementary integrations to evaluate the first few coefficients in (5.28) leads to

$$w_{n+1} = w_n + h\{f_{n+1} - \tfrac{1}{2}\nabla f_{n+1} - \tfrac{1}{12}\nabla^2 f_{n+1} - \tfrac{1}{24}\nabla^3 f_{n+1}$$
$$- \tfrac{19}{720}\nabla^4 f_{n+1} - \tfrac{3}{160}\nabla^5 f_{n+1} - \cdots + b_N\nabla^N f_{n+1}\}. \tag{5.30}$$

Taking $N = 1, 2, 3, 4$ in turn and using equation (3.24) to express the backward differences in terms of function values leads to the following integration schemes:

Second-order, one-step

$$w_{n+1} = w_n + \frac{h}{2}\{f(x_{n+1}, w_{n+1}) + f_n\}, \quad \varepsilon_{n+1} = -\frac{h^3}{12}y^{(3)}(\xi_n). \tag{5.31}$$

Third-order, two-step

$$w_{n+1} = w_n + \frac{h}{12}\{5f(x_{n+1}, w_{n+1}) + 8f_n - f_{n-1}\},$$

$$\varepsilon_{n+1} = -\frac{h^4}{24}y^{(4)}(\xi_n). \tag{5.32}$$

Fourth-order, three-step

$$w_{n+1} = w_n + \frac{h}{24}\{9f(x_{n+1}, w_{n+1}) + 19f_n - 5f_{n-1} + f_{n-2}\},$$

$$\varepsilon_{n+1} = -\tfrac{19}{720}h^5 y^{(5)}(\xi_n). \tag{5.33}$$

Fifth-order, four-step

$$w_{n+1} = w_n + \frac{h}{720}\{251f(x_{n+1}, w_{n+1}) + 646f_n - 264f_{n-1} + 106f_{n-2} - 19f_{n-3}\},$$

$$\varepsilon_{n+1} = -\tfrac{3}{160}h^6 y^{(6)}(\xi_n). \tag{5.34}$$

Higher order schemes can be obtained by taking larger values of N in equation (5.30).

All of the formulae that result from equation (5.26) are called Adams–Moulton formulae. Whereas the Adams–Bashforth schemes were obtained using open integration, and explicit formulae of the form

$$w_{n+1} = F(w_n, w_{n-1}, \cdots) \tag{5.35}$$

were obtained, the Adams–Moulton schemes have been obtained via closed integration and they have the general form

$$w_{n+1} = G(w_{n+1}, w_n, w_{n-1}, \cdots), \tag{5.36}$$

i.e. the Adams–Moulton schemes yield implicit formulae for w_{n+1}. Consequently Adams–Moulton schemes have to be used iteratively to find w_n. The iterative scheme is obtained by expressing (5.36) in the form

$$w_{n+1}^{(k+1)} = G(w_{n+1}^{(k)}, w_n, w_{n-1}, \cdots), \tag{5.37}$$

where $w_{n+1}^{(k)}$ is the kth estimate of w_{n+1}. The convergence of the scheme is discussed later, but when it does converge it yields the solution of (5.36).

All of the schemes (5.31)–(5.34) must normally be applied in the iterative form (5.37), e.g. to find w_{n+1} from (5.32) we express it in the form

$$w_{n+1}^{(k+1)} = w_n + \frac{h}{12}\{5f(x_{n+1}, w_{n+1}^{(k)}) + 8f_n - f_{n-1}\}. \tag{5.38}$$

Equation (5.37) will generate the sequence $w_{n+1}^{(1)}, w_{n+1}^{(2)}, \ldots$ from a starting value or initial guess $w_{n+1}^{(0)}$. This initial guess or prediction can be obtained from an explicit formula of the form (5.35). Thus, in practice, the Adams–Bashforth schemes and Adams–Moulton schemes are used together. The former supplies a predicted solution and the latter corrects or improves this initial prediction. Methods of this type are called predictor–corrector methods.

Predictor–corrector methods, other than the Adams–Bashforth–Moulton type, do exist. A scheme that used to be popular was the Milne–Simpson method. This is described in section 5.1.4.

When a predictor–corrector method is used to solve a differential equation, predictor and corrector formulae of the same order are chosen. For example, one of the most popular predictor–corrector methods uses the fourth-order Adams–Bashforth formula (5.16) for prediction and the fourth-order Adams–Moulton formula (5.33) for correction:

$$w_{n+1}^{(0)} = w_n + \frac{h}{24}\{55f_n - 59f_{n-1} + 37f_{n-2} - 9f_{n-3}\}$$

$$w_{n+1}^{(k+1)} = w_n + \frac{h}{24}\{9f(x_{n+1}, w_{n+1}^{(k)}) + 19f_n - 5f_{n-1} + f_{n-2}\}. \tag{5.39}$$

The overall accuracy of a solution would not be affected by using a predictor formula of lower order than the corrector formula. However this may lead to additional calculations to achieve a specified accuracy; if the predictor solution is not sufficiently accurate, more iterations with the corrector formula may be necessary.

A feature of multi-step methods is that they can only be used when the numerical solution is already available over several consecutive steps. For example, the prediction formula in (5.39) can only be used to calculate w_{n+1} when values of w_n, w_{n-1}, w_{n-2} and w_{n-3} are available. This is necessary for the calculation of f_n, f_{n-1}, f_{n-2} and f_{n-3}. Recalling that we are solving the initial-value problem (5.1) with initial condition of the form (5.2), the only detail available at the start of the problem is $w_0 = y(x_0)$. Consequently a different method must be used to calculate w_1, w_2, and w_3. More generally when using an Nth-order predictor formula a different method must be used to calculate $w_1, w_2, w_3, \ldots, w_{N-1}$. It is therefore necessary to develop starting methods before we can apply the linear multi-step schemes to specific problems.

5.1.3 Starting methods

(i) Taylor series method

One method of initiating the solution of the initial-value problem

$$\frac{dy}{dx} = f(x, y), \quad y(x_0) = y_0$$

in via the Taylor expansion

$$y(x_0 + rh) = y_0 + \frac{rh}{1!}y'_0 + \frac{r^2h^2}{2!}y''_0 + \cdots + \frac{r^kh^k}{k!}y_0^{(k)} + \frac{r^{k+1}h^{k+1}}{(k+1)!}y^{(k+1)}(\xi), \quad (5.40)$$

where $x_0 \leqslant \xi \leqslant x_0 + rh$. The values of y'_0, y''_0, $y'''_0, \ldots$, can be determined in turn by successively differentiating and substituting the values of x_0 and y_0 for x and y. The values of $w_1 \simeq y(x_0 + h)$, $w_2 \simeq y(x_0 + 2h)$, $w_3 \simeq y(x_0 + 3h), \ldots$ can then be obtained by setting $r = 1, 2, 3, \ldots$ in (5.40).

In some problems it is not easy to achieve sufficient accuracy with this method and often a large number of terms is needed in (5.40). This problem is more pronounced with large values of r. Consequently, when more than three starting values are required (for predictor formulae with order greater than four) it is usually preferable to use a new Taylor expansion, based at x_3 for example:

$$y(x_3 + sh) = y_3 + \frac{sh}{1!}y'_3 + \frac{s^2h^2}{2!}y''_3 + \cdots. \quad (5.41)$$

We have already mentioned that the fourth-order scheme (5.39) is suitable for many problems. The starting values for the predictor formula can be obtained from (5.40) with $r = 1, 2, 3$. If there is difficulty in obtaining the required accuracy, better starting values can be calculated using the iterative improvement method described in section 5.1.3(iii).

Example 5.1 Use the Taylor series method to estimate $y(0.1)$, $y(0.2)$ and $y(0.3)$ to three decimal places for the initial-value problem

$$\frac{dy}{dx} = x + y^2, \quad y(0) = 0.$$

From the differential equation $y' = x + y^2$ and the initial condition $y(0) = 0$ (i.e. $x_0 = 0$, $y_0 = 0$) we can determine the values of the derivatives y'_0, $y''_0, \ldots$ in (5.40). Thus by differentiating and substituting we find that

$$
\begin{aligned}
y' &= x + y^2, & y'(0) &= 0 \\
y'' &= 1 + 2yy', & y''(0) &= 1 \\
y^{(3)} &= 2y'^2 + 2yy'', & y^{(3)}(0) &= 0 \\
y^{(4)} &= 6y'y'' + 2yy^{(3)}, & y^{(4)}(0) &= 0 \\
y^{(5)} &= 6y''^2 + 2y'y^{(3)} + 2yy^{(4)}, & y^{(5)}(0) &= 6.
\end{aligned}
$$

Thus, for this problem, the Taylor expansion (5.40) yields

$$y(rh) = \frac{r^2h^2}{2} + \frac{r^5h^5}{20} + \cdots.$$

Setting $h = 0.1$ and $r = 1, 2, 3$ in turn yields the estimates

$$y(0.1) = 0.0050 + 0.0000$$
$$y(0.2) = 0.0200 + 0.0000$$
$$y(0.3) = 0.045 + 0.0001$$

respectively. Thus to three decimal places we have

$$y(0.1) = 0.005, \quad y(0.2) = 0.020, \quad y(0.3) = 0.045.$$

(ii) Picard's method

Here the first-order differential equation (5.1) with initial condition (5.2) is integrated from x_0 to a general point x, giving

$$y(x) = y_0 + \int_{x_0}^{x} f(x, y(x)) dx \qquad (5.42)$$

and successive approximations to $y(x)$ are generated by writing (5.42) in the iterative form

$$y(x) = y_0 + \int_{x_0}^{x} f(x, y^{(k)}(x)) dx. \qquad (5.43)$$

Although we do not discuss the conditions that $f(x, y)$ must satisfy for convergence to be achieved it is clear that, when (5.43) does converge, it yields the solution $y(x)$ of (5.42).

We are familiar with iterative techniques that yield a sequence of numerical approximations to a given problem. The iteration in (5.43) is quite different. From the initial guess $y^{(0)}(x)$, which is normally taken as the initial condition $y_0 = y(x_0)$, it produces a sequence of functions

$$y^{(1)}(x), \quad y^{(2)}(x), \quad y^{(3)}(x), \dots, y^{(k)}(x), \quad y^{(k+1)}(x), \cdots \qquad (5.44)$$

that approximates the solution $y(x)$. We find that the elements of this sequence are finite series and that each stage of the iteration increases the number of terms in the current approximation to $y(x)$. Some stages of the iteration will add more than one term to the current estimate. However, only one additional term, that of the lowest order, should be retained at each stage of the iteration. The other additional terms, called spurious terms, should be discarded.

Picard's method is not always easy to apply. This depends on the form of the integrand $f(x, y)$ in (5.42).

Example 5.2 Repeat Example 5.1 using Picard's method instead of the Taylor series method.

Taking the initial guess as the initial condition we have $y^{(0)}(0) = 0$ and since $f(x, y) = x + y^2$, Picard's method (5.43) yields

$$y^{(1)}(x) = \int_0^x (x + (y^{(0)}(x))^2) dx$$

$$= \frac{x^2}{2}.$$

Proceeding with Picard's method we obtain

$$y^{(2)}(x) = \int_0^x (x + (y^{(1)}(x))^2) dx$$

$$= \int_0^x \left(x + \frac{x^4}{4} \right) dx$$

$$= \frac{x^2}{2} + \frac{x^5}{20},$$

$$y^{(3)}(x) = \int_0^x (x + (y^{(2)}(x))^2)\,dx$$

$$= \int_0^x \left(x + \frac{x^4}{4} + \frac{x^7}{20} + \frac{x^{10}}{400} \right) dx$$

$$= \frac{x^2}{2} + \frac{x^5}{20} + \frac{x^8}{160} + \frac{x^{11}}{4400}.$$

In this last stage of the iteration we notice that $y^{(3)}(x)$ contains two additional terms compared to $y^{(2)}(x)$. We only retain the lowest order term and discard the higher order, spurious term. Thus we have

$$y^{(3)}(x) = \frac{x^2}{2} + \frac{x^5}{20} + \frac{x^8}{160}$$

and substituting $x = 0.1$, 0.2 and 0.3 in turn leads to the estimates

$$y(0.1) = 0.00500, \quad y(0.2) = 0.02001, \quad y(0.3) = 0.04512.$$

Examination of the magnitude of the last term, $x^8/160$, in $y^{(3)}(x)$, for each value of x, indicates that these results are accurate to five decimal places.

For the initial-value problem specified in Example 5.1 Picard's method has been a little easier to use than the Taylor series method. This depends on the form of $f(x, y)$ and is certainly not always the case. Notice that

$$y(rh) \simeq \frac{r^2 h^2}{2} + \frac{r^5 h^5}{20}$$

from the Taylor series method (Example 5.1) agrees exactly with

$$y^{(2)}(x) = \frac{x^2}{2} + \frac{x^5}{20}$$

from Picard's method (Example 5.2) since $x = x_0 + rh, x_0 = 0$. The additional term in $y^{(3)}(x)$ gave the increased accuracy when using Picard's method. With some extra effort this additional term could also have been included in the Taylor series approach.

(iii) Iterative improvement

With some problems the Taylor series method and Picard's method are difficult or laborious to apply for finding starting values to a required accuracy. An alternative approach is to use one of these methods to obtain initial estimates for the starting values. Then an iterative method can be used to obtain better estimates for these starting values and achieve the required accuracy. We describe a suitable iterative method for improving the starting values.

Suppose that we are planning to use the fourth-order predictor–corrector method (5.39). Then from the initial condition $y(x_0) = y_0$ we must obtain starting values for $w_1 \simeq y_1$, $w_2 \simeq y_2$ and $w_3 \simeq y_3$. Initial estimates for these starting values can be obtained using Taylor's series method or Picard's method. From those initial estimates we can obtain the approximations $f_1 \simeq f(x_1, w_1)$, $f_2 \simeq f(x_2, w_2)$ and $f_3 \simeq f(x_3, w_3)$. Then the integrand $f(x, y)$ in (5.42) can be approximated by the cubic polynomial passing through (x_0, f_0), (x_1, f_1), (x_2, f_2) and (x_3, f_3). Using the Gregory–Newton forward version of this polynomial (equation (3.30)) we obtain

$$y_r = y_0 + \int_0^1 \left\{ f_0 + r\Delta f_0 + \frac{r(r-1)}{2!} \Delta^2 f_0 + \frac{r(r-1)(r-2)}{3!} \Delta^3 f_0 \right\} h\, dr + \varepsilon_r \qquad (5.45)$$

where $x = x_0 + rh$, $y_r = y(x_0 + rh)$ and ε_r is the error associated with the cubic approximation to the integrand. Carrying out the elementary integrations in (5.45) and setting $r = 1, 2, 3$ in turn leads to new estimates for y_1, y_2, y_3:

$$w_1 = w_0 + h\{ f_0 + \tfrac{1}{2}\Delta f_0 - \tfrac{1}{12}\Delta^2 f_0 + \tfrac{1}{24}\Delta^3 f_0 \}, \quad \varepsilon_1 = -\tfrac{19}{720} h^5 y^{(5)}(\xi) \qquad (5.46)$$

$$w_2 = w_0 + h\{ 2f_0 + 2\Delta f_0 + \tfrac{1}{3}\Delta^2 f_0 \}, \quad \varepsilon_2 = -\frac{h^5}{90} y^{(5)}(\xi) \qquad (5.47)$$

$$w_3 = w_0 + h\{ 3f_0 + \tfrac{9}{2}\Delta f_0 + \tfrac{9}{4}\Delta^2 f_0 + \tfrac{3}{8}\Delta^3 f_0 \}, \quad \varepsilon_3 = -\tfrac{3}{80} h^5 y^{(5)}(\xi). \qquad (5.48)$$

The error terms can be obtained by considering the leading contribution to the terms omitted from the Gregory–Newton formula, as described in the derivations of the linear multi-step formulae. Using equation (3.23) to express the finite differences in terms of function values we obtain

$$w_1 = w_0 + \frac{h}{24} \{ 9f_0 + 19f_1 - 5f_2 + f_3 \} \qquad (5.49)$$

$$w_2 = w_0 + \frac{h}{3} \{ f_0 + 4f_1 + f_2 \} \qquad (5.50)$$

$$w_3 = w_0 + \frac{3h}{8} \{ f_0 + 3f_1 + 3f_2 + f_3 \}. \qquad (5.51)$$

Since f_0 and approximate values for f_1, f_2 and f_3 are available, these formulae can be used to obtain new estimates for w_1, w_2 and w_3. If these values are not in agreement with the initial estimates, to the required accuracy, f_1, f_2 and f_3 should be recalculated. Equations (5.49), (5.50) and (5.51) are then used again to obtain another set of estimates for w_1, w_2 and w_3. Proceeding in this manner we generate the finite sequences

$$w_1^{(0)}, \quad w_1^{(1)}, \quad w_1^{(2)}, \ldots w_1^{(N)}$$
$$w_2^{(0)}, \quad w_2^{(1)}, \quad w_2^{(3)}, \ldots, w_2^{(N)}$$
$$w_3^{(0)}, \quad w_3^{(1)}, \quad w_3^{(2)}, \ldots, w_3^{(N)},$$

where the value of N is such that

$$\left| w_j^{(N-1)} - w_j^{(N)} \right| < \varepsilon \quad \text{for } j = 1, 2, 3 \qquad (5.52)$$

and ε is some prescribed tolerance that yields the required accuracy, i.e. the iteration is continued until two successive members of each sequence agree to the required number of decimal places.

In summary, the iteration described above can be represented by writing equations (5.49), (5.50) and (5.51) in the form

$$w_1^{(k+1)} = w_0 + \frac{h}{24} \{ 9f_0 + 19f_1^{(k)} - 5f_2^{(k)} + f_3^{(k)} \}$$

$$w_2^{(k+1)} = w_0 + \frac{h}{3}\{f_0 + 4f_1^{(k)} + f_1^{(k)}\} \tag{5.53}$$

$$w_3^{(k+1)} = w_0 + \frac{3h}{8}\{f_0 + 3f_1^{(k)} + 3f_2^{(k)} + f_3^{(k)}\}.$$

However, the convergence of the iteration may be improved by using the latest estimates of w_1, w_2 and w_3 as soon as they are available, i.e. instead of (5.53) use the similar scheme

$$w_1^{(k+1)} = w_0 + \frac{h}{24}\{9f_0 + 19f_1^{(k)} - 5f_2^{(k)} + f_3^{(k)}\}$$

$$w_2^{(k+1)} = w_0 + \frac{h}{3}\{f_0 + 4f_1^{(k+1)} + f_2^{(k)}\} \tag{5.54}$$

$$w_3^{(k+1)} = w_0 + \frac{3h}{8}\{f_0 + 3f_1^{(k+1)} + 3f_2^{(k+1)} + f_3^{(k)}\}.$$

We have derived these iterative formulae to improve the starting values for a fourth-order predictor–corrector method. In a similar manner, iterative formulae can be derived for predictor–corrector methods with different order.

Example 5.3 For the initial-value problem

$$\frac{dy}{dx} = x + y^2, \quad y(0) = 0$$

use a starting method to estimate $y(0.2)$, $y(0.4)$, $y(0.6)$ and use a fourth-order predictor–corrector method to estimate $y(0.8)$.

In this example we demonstrate the use of the iterative improvement technique and the fourth-order predictor–corrector formulae (5.39). Consequently we need not be too concerned about the accuracy of the initial estimates for the starting values. The initial-value problem in this example was also examined in Examples 5.1 and 5.2, but different function values were required.

Following Example 5.1 the Taylor series method leads to

$$y(rh) = \frac{r^2 h^2}{2} + \frac{r^5 h^5}{20} + \cdots$$

and for this problem we take $h = 0.2$. Working to five decimal places throughout and taking $r = 1,2,3$ in turn leads us to the initial estimates

$$w_1^{(0)}(0.2) = 0.02002, \quad w_2^{(0)}(0.4) = 0.08051, \quad w_3^{(0)}(0.6) = 0.18389$$

for the starting values $y(0.2)$, $y(0.4)$ and $y(0.6)$. At this stage it is not necessary to check the accuracy since that will be part of the iterative improvement scheme. From these starting values we determine initial estimates for $f_1 = f(0.2, y(0.2))$, $f_2 = f(0.4, y(0.4))$ and $f_3 = f(0.6, y(0.6))$. In this problem, $f(x, y) = x + y^2$ and therefore

$$f_1^{(0)} = 0.20040, \quad f_2^{(0)} = 0.40648, \quad f_3^{(0)} = 0.63382.$$

Substituting these values and $w_0 = 0$, $f_0 = 0$ into the first equation of (5.54) yields

$$w_1^{(1)} = w_0 + \frac{h}{24}\{9f_0 + 19f_1^{(0)} - 5f_3^{(0)}\}$$

$$= 0.02008.$$

Hence $f_1^{(1)} = 0.20040$ and the second equation of (5.54) yields

$$w_2^{(1)} = w_0 + \frac{h}{3}\{f_0 + 4f_1^{(1)} + f_2^{(0)}\}$$

$$= 0.08054.$$

Hence $f_2^{(1)} = 0.40649$ and the third equation of (5.54) yields

$$w_3^{(1)} = w_0 + \frac{3h}{8}\{f_0 + 3f_1^{(1)} + 3f_2^{(1)} + f_3^{(0)}\}$$

$$= 0.18409.$$

Proceeding in this manner we calculate in turn

$$f_3^{(1)} = 0.63389$$

$$w_1^{(2)} = w_0 + \frac{h}{24}\{9f_0 + 19f_1^{(1)} - 5f_2^{(1)} + f_3^{(1)}\}$$

$$= 0.02008 \; (= w_1^{(1)})$$

$$f_1^{(2)} = 0.20040 \; (= f_1^{(1)})$$

$$w_2^{(2)} = w_0 + \frac{h}{3}\{f_0 + 4f_1^{(2)} + f_2^{(1)}\}$$

$$= 0.08054 \; (= w_2^{(1)})$$

$$f_2^{(2)} = 0.40649 \; (= f_2^{(1)})$$

$$w_3^{(2)} = w_0 + \frac{3h}{8}\{f_0 + 3f_1^{(2)} + 3f_2^{(2)} + f_3^{(1)}\}$$

$$= 0.18409 \; (= w_3^{(1)}).$$

The scheme has converged to the starting values

$$w_1 = 0.02008, \quad w_2 = 0.08054, \quad w_3 = 0.18409$$

and we now use the fourth-order predictor–corrector method (5.39) to estimate $w_4 \simeq y(0.8)$. From the predictor formula we obtain

$$w_4^{(0)} = w_3 + \frac{h}{24}\{55f_3 - 59f_2 + 37f_1 - 9f_0\}$$

$$= 0.33656.$$

Hence $f_4^{(0)} = 0.91327$ and the corrector formula gives

$$w_4^{(1)} = w_3 + \frac{h}{24}\{9f_4^{(0)} + 19f_3 - 5f_2 + f_1\}$$

$$= 0.33768.$$

Since this does not agree with $w_4^{(0)}$ we calculate $f_4^{(1)} = 0.91403$ and use the corrector formula again:

$$w_4^{(2)} = w_3 + \frac{h}{24} \{9 f_4^{(1)} + 19 f_3 - 5 f_2 + f_1\}$$

$$= 0.33774.$$

At this stage we observe that $w_4^{(1)}$ and $w_4^{(2)}$ agree when rounded to four decimal places, i.e.

$$y(0.8) \simeq w_4 = 0.3377.$$

The predictor–corrector scheme can be used to advance the scheme further. For example, to estimate $y_5 = y(1.0)$, $y_6 = y(1.2)$, and so on we use (5.39) with $n = 4, 5$, and so on in turn.

Algorithm 5.1. The fourth-order predictor–corrector method with iterative improvement

```
read(x0, w0, h, n, Tol);
read(w1, w2, w3);
x1:= x0 + h; x2:= x1 + h; x3:= x2 + h; x4:= x3 + h;
f0:= f(x0, w0); f1:= f(x1, w1); f2:= f(x2, w2); f3:= f(x3, x3);
(* apply iterative improvement*)
Loop:= true;
WHILE Loop DO
   BEGIN
   w101d:= w1; w201d:= w2; w301d:= w3;
   w1:= w0 + h*(9* f0 + 19* f1 − 5* f2 + f3)/24;
   f1:= f(x1, w1);
   w2:= w0 + h*(f0 + 4* f1 + f2)/3;
   f2:= f(x₂, w2);
   w3:= w0 + 3*h*(f:= + 3* f1 + 3* f2 + f3)/8;
   f3:= f(x3, w3);
   Loop:= (abs(w1-w101d) > Tol) OR (abs(w2-w201d) > Tol) OR
          (abs(w3-w301d) > Tol)
   END;
(*predictor–corrector sequence*)
FOR i:= 4 TO n DO
   BEGIN
   w4:= w3 + h*(55* f3 − 59* f2 + 32* f1 − 9* f0);
   REPEAT
      f4:= f(x4, w4); w401d:= w4;
      w4:= w3 + h*(9* f4 + 19* f3 − 5* f2 + f1)/24
   UNTIL abs(w4 − w401d) < Tol;
   write(x4, w4);
      x4:= x4 + h; w3:= w4;
   f0:= f1; f1:= f2; f2:= f3; f3:= f4
END.
```

5.1.4 More general multi-step methods

The numerical integration formulae that we have derived in the previous sections have been obtained by integrating the differential equation (5.1) over the interval $[x_n, x_{n+1}]$. This leads to equation (5.7) and the integrand $f(x, y)$ was approximated by polynomial functions that passed through some of the previous data points (x_n, f_n), $(x_{n-1}, f_{n-1}), \ldots, (x_0, f_0)$. A more general approach is to integrate the differential equation (5.1) over the wider interval $[x_{n-p}, x_{n-1}]$, where $p \leqslant N \leqslant n$ and N is the degree of the polynomial approximation to the integrand $f(x, y)$ (see equations (5.8), (5.26)). With this approach, instead of equation (5.7), we obtain

$$y_{n+1} = y_{n-p} + \int_{x_{n-p}}^{x_{n+1}} f(x, y)\, dx. \tag{5.55}$$

If the integrand is approximated by the Nth-degree Gregory–Newton backward formula based at x_n, as in equation (5.8), we obtain predictor formulae of the form

$$w_{n+1} = w_{n-p} + h \sum_{k=0}^{N} a_k \nabla^k f(x_n, w_n), \tag{5.56}$$

where

$$a_k = \int_{-p}^{1} \frac{r(r+1)\cdots(r+k-1)}{k!} \tag{5.57}$$

and r is defined by $x = x_n + rh$. Corrector formulae are obtained by approximating $f(x, y)$ by the Nth-degree Gregory–Newton backward formulae based at x_{n+1}, as in equation (5.26). They take the form

$$w_{n+1} = w_{n-p} + h \sum_{k=0}^{N} b_k \nabla^k f(x_{n+1}, w_{n+1}), \tag{5.58}$$

where

$$b_k = \int_{-p-1}^{0} \frac{s(s+1)\cdots(s+k-1)}{k!}\, ds \tag{5.59}$$

and $x = x_{n+1} + sh$.

One of the most common schemes of this type is the Milne–Simpson method. Milne's predictor formula is obtained by approximating the integrand $f(x, y)$ over $[x_{n-3}, x_{n+1}]$ by the quadratic passing through (x_i, f_i) for $i = n$, $n-1$, $n-2$ and integrating over $[x_{n-3}, x_{n+1}]$, i.e. $p = 3$ in (5.56), (5.57). This leads to Milne's open formula

$$w_{n+1} = w_{n-3} + \frac{4h}{3}(2f_n - f_{n-1} + 2f_{n-2}), \quad \varepsilon_{n+1} = \tfrac{14}{15} h^5 y^{(5)}(\xi_n). \tag{5.60}$$

Simpson's corrector formula is obtained by approximating the integrand $f(x, y)$ over $[x_{n-1}, x_{n+1}]$ by the quadratic passing through (x_i, f_i) for $i = n+1$, n, $n-1$ and integrating over $[x_{n-1}, x_{n+1}]$. This leads to Simpson's closed formula

$$w_{n+1} = w_{n-1} + \frac{h}{3}\{f(x_{n+1}, w_{n+1}) + 4f_n + f_{n-1}\}, \quad \varepsilon_{n+1} - \frac{h^5}{90} y^{(5)}(\xi_n). \tag{5.61}$$

Although the Milne–Simpson method used to be popular it can give rise to large errors when integrating over many steps. The Adams–Bashforth–Moulton schemes do not suffer from this difficulty and are therefore preferred. This problem of whether or not a given solution procedure remains stable when integrating over several steps is discussed in section 5.1.5.

Even more general formulae than those arising from equation (5.56) and (5.58) can be derived by other methods.

5.1.5 Accuracy of multi-step formulae

We have derived several formulae for obtaining the approximate numerical solution of the differential equation (5.1). In this section we examine the errors associated with the approximations. We identify three possible sources of error: local truncation error, error associated with the iterative solution of the corrector formula and error propagation through the step-wise integration.

(i) Estimation of truncation error

The local truncation errors have already been given with each of the predictor and corrector formulae. However these are of little use for obtaining numerical estimates of the truncation error. Here we describe one method for obtaining an estimate of the truncation error. The method is demonstrated for the fourth-order scheme (5.39).

We will examine the error associated with the approximation w_{n+1} for $y_{n+1} = y(x_{n+1})$. Since we are dealing with error propagation as a separate issue we will assume that the errors in previously calculated estimates are negligible. This is reasonable provided that the error is controlled at each step and that the solution scheme does not propagate small errors into large errors at a later stage of the step-wise integration. With this assumption we may write

$$w_i \simeq y_i, \quad f(x_i, w_i) \simeq f(x_i, y_i) \quad \text{for } i = 0, 1, \ldots, n.$$

The exact solution y_{n+1}, of the differential equation, satisfies the two equations

$$y_{n+1} = y_n + \frac{h}{24}(55 f_n - 59 f_{n-1} + 37 f_{n-2} - 9 f_{n-3}) + \tfrac{251}{720} h^5 y^{(5)}(\alpha_n) \tag{5.62}$$

and

$$y_{n+1} = y_n + \frac{h}{24}(9 f_{n+1} + 19 f_n - 5 f_{n-1} + f_{n-2}) - \tfrac{19}{720} h^5 y^{(5)}(\beta_n) \tag{5.63}$$

which follow directly from (5.16) and (5.33) by adding in the error terms. Recall that the w-notation in (5.16), (5.33) was introduced to emphasize that these formulae were obtained by truncating an infinite series, i.e. neglecting the truncation error. In the error terms we have changed the notation from ξ_n to α_n for the predictor formula and from ξ_n to β_n for the corrector formula. This is to emphasize that the derivatives are evaluated at different values of x in general.

Forming the difference of equations (5.16) and (5.62) gives

$$y_{n+1} - w_{n+1}^{(0)} = \tfrac{251}{720} h^5 y^{(5)}(\alpha_n), \tag{5.64}$$

where we have used $w_{n+1}^{(0)}$ rather than w_{n+1} on the left-hand side of (5.16) to emphasize that (5.16) yields the initial or predicted value for w_{n+1}. Similarly, forming the difference between equations (5.33) and (5.63) gives

$$y_{n+1} - w_{n+1} = \frac{9h}{24}\{f(x_{n+1}, y_{n+1}) - f(x_{n+1}, w_{n+1})\} - \tfrac{19}{720}h^5 y^{(5)}(\beta_n)$$

and using the mean-value theorem

$$y_{n+1} - w_{n+1} = \frac{3h}{8}(y_{n+1} - w_{n+1})f_y(x_{n+1}, \eta_{n+1}) - \tfrac{19}{720}h^5 y^{(5)}(\beta_n) \qquad (5.65)$$

where f_y is the partial derivative of $f = f(x, y)$ with respect to y and η_{n+1} lies between y_{n+1} and w_{n+1}. Equation (5.65) can be written

$$(y_{n+1} - w_{n+1})\left\{1 - \frac{3h}{8}f_y(x_{n+1}, \eta_{n+1})\right\} = \tfrac{19}{720}h^5 y^{(5)}(\beta_n)$$

and assuming that h is chosen sufficiently small to ensure that

$$\frac{3h}{8}|f_y(x_{n+1}, \eta_{n+1})| \ll 1 \qquad (5.66)$$

(the symbol '$\ll$' means 'very much less than') we have

$$y_{n+1} - w_{n+1} = -\tfrac{19}{720}h^5 y^{(5)}(\beta_n). \qquad (5.67)$$

In section 5.1.5 we will see that such a choice of h is also necessary to guarantee convergence of the corrector formula.

To obtain an estimate of the truncation error we need to eliminate the derivative between equations (5.64) and (5.67). However these are evaluated at $x = \alpha_n$ and $x = \beta_n$ and therefore, in general,

$$y^{(5)}(\alpha_n) \neq y^{(5)}(\beta_n).$$

To overcome this difficulty we make the assumption that $y^{(5)}(x)$ does not vary strongly, i.e. we assume that

$$y^{(5)}(\alpha_n) \simeq y^{(5)}(\beta_n). \qquad (5.68)$$

With this assumption we eliminate the fifth derivative between equations (5.64) and (5.67) to obtain

$$y_{n+1} - w_{n+1} \simeq -\tfrac{19}{251}(y_{n+1} - w_{n+1}^{(0)}).$$

Subtracting and adding w_{n+1} inside the brackets of this equation and rearranging gives

$$(y_{n+1} - w_{n+1})(1 + \tfrac{19}{251}) \simeq -\tfrac{19}{251}(w_{n+1} - w_{n+1}^{(0)}),$$

which can be written

$$y_{n+1} - w_{n+1} \simeq -\tfrac{19}{720}(w_{n+1} - w_{n+1}^{(0)}). \qquad (5.69)$$

The right-hand side of the equation (5.69) can be evaluated. It is a constant multiplied by the difference between the solution of the corrector formula and the solution of the predictor formula. In this way the error in the converged solution of the corrector formula can be estimated. The reliability of this estimate depends on the validity of the assumptions made. In particular we assumed that errors propagated from previous calculations are negligible. We also assumed that $y^{(5)}(x)$ does not vary strongly. An indication of the validity of this assumption can be obtained by examining the variation in the first neglected differences. For the fourth-order scheme these are the fourth differences $\nabla^4 f_n$.

The error estimate given by equation (5.69) was derived for the fourthorder predictor–corrector scheme (5.39). Similar estimates, involving the difference between the solution of the corrector formula and the solution of the predictor formula, can be derived for schemes of different order.

(ii) Convergence of the corrector formula

Again we concentrate on the fourth-order predictor–corrector scheme (5.39) and investigate the conditions under which the corrector formula converges. By convergence we mean that the difference between $w_{n+1}^{(k+1)}$ and w_{n+1}, the solution of (5.33), can be made arbitrarily small by taking a sufficiently large value of k. Thus to examine convergence we form the difference of the iterative corrector formula in (5.39) and the converged corrector formula (5.33). This gives

$$w_{n+1} - w_{n+1}^{(k+1)} = \frac{3h}{8}\{f(x_{n+1}, w_{n+1}) - f(x_{n+1}, w_{n+1}^{(k)})\} \tag{5.70}$$

and we see that convergence depends on the variation of $f(x, y)$ with respect to y. This becomes more evident when we use the mean-value theorem, which states that there exists $\eta_{n+1}^{(k)}$, lying between w_{n+1} and $w_{n+1}^{(k)}$, such that

$$f_y(\eta_{n+1}^{(k)}) = \frac{f(x_{n+1}, w_{n+1}) - f(x_{n+1}, w_{n+1}^{(k)})}{(w_{n+1} - w_{n+1}^{(k)})}.$$

Then (5.70) may be written

$$w_{n+1} - w_{n+1}^{(k+1)} = \frac{3h}{8} f_y(x_{n+1}, \eta_{n+1}^{(k)})(w_{n+1} - w_{n+1}^{(k)}). \tag{5.71}$$

At this stage we define the convergence factor ρ_{n+1} by the equation

$$w_{n+1} - w_{n+1}^{(k+1)} = \rho_{n+1}(w_{n+1} - w_{n+1}^{(k)}) \tag{5.72}$$

and we see that convergence occurs when

$$|\rho_{n+1}| < 1. \tag{5.73}$$

Thus, from (5.71) and (5.72), for the fourth-order Adams–Moulton scheme

$$\rho_n \simeq \frac{3h}{8} f_y(x_n, y_n). \tag{5.74}$$

We see that convergence is dependent on the step length h, which we have control over, and on the rate of change of $f(x, y)$ with respect to y, which we have no control over. Intuitively it should not be too surprising that convergence depends on these two factors. One would expect the step length to have some influence, and it is the changes in $f(x, y)$ with respect to y that lead to the different elements of the sequence $w_{n+1}^{(0)}$, $w_{n+1}^{(1)}$, $w_{n+1}^{(2)}, \ldots$ (see (5.39)). In particular, equations (5.73) and (5.74) imply that the fourth-order Adams–Moulton scheme converges if the step length h is chosen to satisfy the inequality

$$h < \frac{8}{3|f_y(x, y)|} \tag{5.75}$$

over the range of interest.

With this approach, similar convergence factors can be derived for other corrector formulae. It can be shown that each of (5.31), (5.32) and (5.34) converges if the step length h is chosen to satisfy

$$h < \frac{2}{|f_y(x, y)|}, \tag{5.76}$$

$$h < \frac{12}{5|f_y(x, y)|}, \tag{5.77}$$

$$h < \frac{720}{251|f_y(x, y)|} \tag{5.78}$$

respectively over the region of interest.

(iii) Error propagation—stability

With modern computational aids it is not unusual to undertake problems in which a differential equation has to be integrated over a very large number of step lengths. Errors are introduced into the numerical solution at all stages of the calculation. Errors that are introduced at an early stage of the integration may die away or may be propagated with increasing magnitude through the step-wise integration. Whether the error propagates or dies away can depend on the differential equation, the boundary conditions or the numerical procedure being used for obtaining the solution. In particular we show that certain finite difference equations of the form (5.58) will diverge from the true solution of the differential equation even if exact arithmetic could be used to compute the solution.

Mathematically, the word 'stability' is used in many different contexts. Here we mainly concentrate on the stability of the numerical procedures. Thus given an initial-value problem, we wish to determine those schemes which may diverge from the solution. However, first we draw attention to the fact that instability can be inherent in the initial-value problem itself, regardless of which numerical scheme is used to obtain the solution. We demonstrate this by considering the differential equation

$$y'' - 5y' - 14y = 0$$

with the initial conditions

$$y(0) = 1, \quad y'(0) = -2.$$

It is easily checked that the solution to this initial-value problem is $y = e^{-2x}$. To examine the stability of this solution we allow one of the initial conditions to change by a small amount ε. Thus we examine the solution of the same differential equation with initial conditions

$$y(0) = 1 + \varepsilon, \quad y'(0) = -2.$$

Again, it is easily checked that the solution to this initial-value problem is

$$y(x) = (1 + \tfrac{7}{9}\varepsilon)e^{-2x} + \tfrac{2}{9}\varepsilon e^{7x}.$$

For any $\varepsilon > 0$, this solution tends to infinity as $x \to \infty$. Therefore we see that a small change in the initial condition produces a large change in the solution, i.e. the solution $y = e^{-2x}$ is an unstable solution. Such problems are often termed ill-conditioned and are difficult to solve numerically since truncation error and round-off error have the same effect as changing the boundary conditions.

To investigate the stability of the numerical schemes we will initially consider Simpson's corrector formula (5.61) applied to the initial-value problem

$$y' = cy, \quad y(x_0) = y_0. \tag{5.79}$$

For this problem we have $f(x, y) = cy$ and substituting into (5.61) gives

$$\left\{1 - \frac{ch}{3}\right\}w_{n+1} - 4\frac{ch}{3}w_n - \left\{1 + \frac{ch}{3}\right\}w_{n-1} = 0. \tag{5.80}$$

This is a second-order linear difference equation and has solution (see Appendix)

$$w_n = a_0\lambda_0^n + a_1\lambda_1^n \tag{5.81}$$

where λ_0 and λ_1 are the roots of the auxiliary equation

$$\left\{1 - \frac{ch}{3}\right\}\lambda^2 - 4\frac{ch}{3}\lambda - \left\{1 + \frac{ch}{3}\right\} = 0. \tag{5.82}$$

The roots of this quadratic are

$$\lambda = \left[\frac{2ch}{3} \pm \left\{1 + \frac{c^2h^2}{3}\right\}^{1/2}\right]\left\{1 - \frac{ch}{3}\right\}^{-1}$$

and carrying out two binomial expansions leads to

$$\lambda_0 = 1 + ch + \cdots, \quad \lambda_1 = -1 + \frac{ch}{3} + \cdots. \tag{5.83}$$

Substituting (5.83) into (5.81) we obtain

$$w_n = a_0(1 + ch + \cdots)^{(x_n - x_0)/h} + a_1(-1)^n(1 - ch/3 + \cdots)^{(x_n - x_0)/h} \tag{5.84}$$

where we have replaced n in (5.81) using $x = x_0 + nh$. Provided h is small (5.84) yields

$$w_n \simeq a_0 e^{c(x_n - x_0)} + a_1(-1)^n e^{-c(x_n - x_0)/3}. \tag{5.85}$$

From (5.79), by elementary integration we see that the exact solution of the initial-value problem is

$$y = y_0 e^{c(x - x_0)} \tag{5.86a}$$

so that

$$y_n = y_0 e^{c(x_n - x_0)} \tag{5.86b}$$

and the first term on the right-hand side of (5.85) corresponds to this solution. However, from (5.85), we see that Simpson's corrector formula generates the solution

$$a_1(-1)^n e^{-c(x_n - x_0)/3} \tag{5.87}$$

in addition to the component that approximates the true solution. This additional or spurious solution causes the numerical scheme to become unstable when c is negative. Consequently Simpson's corrector formula is unsuitable for obtaining the solution of (5.79) when $c < 0$.

The underlying reason for instability that we have demonstrated with Simpson's corrector formula is as follows. The differential equation in (5.79) is first order and consequently has only one independent solution. The approximation of this differential equation by Simpson's corrector formula leads to the second-order difference equation (5.80) which has two independent solutions. Only one of these solutions approximates the exact solution of the differential equation and the remaining spurious solution can cause the error to grow exponentially. More generally, when a differential equation is approximated by a difference equation and the degree of the difference equation exceeds the degree of the differential equation, the solution of the difference equation will contain spurious solutions. These spurious solutions are usually called parasitic solutions. Some parasitic solutions are troublesome and grow exponentially through the step-wise integration, i.e. the solution becomes unstable, whereas others cause no problems. It would not be sensible to eliminate the troublesome parasitic solutions by choosing numerical schemes that have no parasitic solutions at all. This approach would limit the order of the numerical procedures being used and the parasitic solutions would be removed at the expense of increasing the truncation errors. Instead we identify those numerical procedures whose parasitic solutions are well behaved and do not cause the numerical solution to become unstable.

We have only investigated the stability of Simpson's corrector formula applied to a simple initial-value problem. This leads to the linear difference equation (5.80) which was easily solved. With other initial-value problems, more complex forms of $f(x, y)$ may lead to a difference equation that is less easy or impossible to solve, particularly if it is non-linear. Other integration formulae can be considerably more difficult than Simpson's rule to analyse using the method illustrated above. In particular this method investigates stability using a finite step length h. This is perhaps the most obvious approach since it is a finite step length that is used in a practical application of an integration scheme. However the difficult stability analysis forces one to consider stability theory in the limit as $h \to 0$.

By appropriate choice of $\alpha_i (i = -1(1)N)$ each of the formulae (5.56) and (5.58) can be expressed in the form

$$w_{n+1} = w_{n-p} + h \sum_{i=-1}^{N} \alpha_i f_{n-i}. \tag{5.88}$$

Since $f = f(x, y)$ and $p \leqslant N \leqslant n$, this difference equation has order (the difference between the highest and lowest suffices)

$$(n + 1) - (n - N) = N + 1. \tag{5.89}$$

This becomes more obvious when we apply the scheme (5.88) to the initial-value problem (5.79). This leads to the linear difference equation

$$(1 - \alpha_{-1} ch) w_{n+1} = w_{n-p} + ch(\alpha_0 w_n + \alpha_1 w_{n-1} + \cdots + \alpha_N w_{n-N}). \tag{5.90}$$

The solution of this equation can be expressed in the form (see Appendix)

$$w_n = \sum_{i=0}^{N} a_i \lambda_i^n \tag{5.91}$$

provided that $\lambda_i (i = 0(1)N)$ are the distinct roots of the polynomial

$$(1 - \alpha_{-1}ch)\lambda^{N+1} - ch(\alpha_0\lambda^N + \alpha_1\lambda^{N-1} + \cdots + \alpha_N) - \lambda^{N-p} = 0. \qquad (5.92)$$

We now consider the limiting case as $h \to 0$. Equation (5.92) reduces to the form

$$\lambda^{N-p}(\lambda^{p+1} - 1) = 0. \qquad (5.93)$$

Although equations (5.90)–(5.92) only apply to the initial-value problem (5.79), the limiting equation (5.93) is valid for more general initial-value problems.

Equation (5.93) has the root $\lambda = 0$ with multiplicity $N - p$. The other roots are the $p + 1$ roots of unity lying on the unit circle $|\lambda| = 1$ in the complex plane. When h takes a small but finite value, equation (5.92) has $N - p$ roots close to the origin in the complex plane and $p + 1$ roots close to the unit circle. One of these $N + 1$ roots corresponds to the true solution of the first-order differential equation. This is the root that approaches $\lambda = 1$, on the real axis, as $h \to 0$. The other N roots lead to parasitic solutions. Referring to (5.91) we see that $N - p$ of these parasitic solutions cause no problems as n increases through the step-wise integration. However, the p parasitic solutions deriving from those roots that lie close to the unit circle will cause the numerical solution to diverge if one or more of these roots lie outside the unit circle, i.e. $|\lambda_i| > 1$.

For Simpson's corrector formula (5.61) we have $p = 1$ (compare (5.61) and (5.88)), i.e. there is one parasitic solution which may cause the numerical solution to diverge. Whether it does cause trouble or not depends on the form of $f(x, y)$ in the differential equation. For the initial-value problem (5.79) we found that this depended on the sign of c, i.e. the sign of c determined whether the root λ_i (see (5.83)) was inside or outside the unit circle.

In general it is not always possible to determine whether the p roots close to the unit circle are inside or outside the unit circle. However, it is possible to select integration schemes for which $p = 0$. Note that the polynomial equation (5.93) then takes the form

$$\lambda^N(\lambda - 1) = 0. \qquad (5.94)$$

All of the Adams–Bashforth and Adams–Moulton methods fall into this category and therefore give no stability problems. However Simpson's corrector formula has $p = 1$ and (5.93) takes the form

$$\lambda^2 - 1 = 0 \qquad (5.95)$$

(see also (5.82) with $h = 0$). Thus, as we have seen, the root close to $\lambda = -1$ can lead to a diverging parasitic solution. Consequently it is better to use one of the Adams' formulae rather than Simpson's formula.

The terms 'stable' and 'strongly stable' are frequently used when discussing the stability of numerical schemes for integrating differential equations. We describe this terminology for the more general linear multi-step method

$$w_{n+1} = \sum_{i=0}^{p} \beta_i w_{n-1} + h \sum_{i=-1}^{N} \alpha_i f_{n-i} \qquad (5.96)$$

which includes schemes of the form (5.88) and consequently all of the methods that we have described. If (5.96) is applied to the initial-value problem (5.79) we obtain the linear difference equation

$$w_{n+1} = \sum_{0}^{p} \beta_i w_{n-i} + hc \sum_{i=-1}^{N} \alpha_i w_{n-i} \qquad (5.97)$$

and, as for (5.88), this equation has order $N + 1$. Recalling that $p \leqslant N$, equation (5.97) can be written in the form

$$(1 - \alpha_{-1}hc)w_{n+1} = \sum_{i=0}^{p} (\beta_i + hc\alpha_i)w_{n-i} + \sum_{i=p+1}^{N} hc\alpha_i w_{n-i} \qquad (5.98)$$

and the solution takes the form given in (5.91) provided that $\lambda_i (i = 0(1)N)$ are the $N + 1$ distinct roots of the polynomial (see the Appendix)

$$(1 - \alpha_{-1}hc)\lambda^{n+1} = \sum_{i=0}^{p} (\beta_i + hc\alpha_i)\lambda^{n-i} + \sum_{i=p+1}^{N} hc\alpha_i \lambda^{n-i},$$

or, dividing through by $\lambda^{n-N} (p \leqslant N \leqslant n)$,

$$(1 - \alpha_{-1}hc)\lambda^{N+1} = \sum_{i=0}^{p} (\beta_i + hc\alpha_i)\lambda^{N-i} + \sum_{i=p+1}^{N} hc\alpha_i \lambda^{N-i}. \qquad (5.99)$$

As with (5.92), we consider the limiting case as $h \to 0$. Then equation (5.99) reduces to the form

$$\lambda^{N+1} - \beta_0\lambda^N - \beta_1\lambda^{N-1} - \cdots - \beta_p\lambda^{N-p} = 0. \qquad (5.100)$$

Note that this reduces to (5.93) when $\beta_0 = \beta_1 = \cdots = \beta_{p-1} = 0$ and $\beta_p = 1$. The polynomial in (5.100) is sometimes referred to as the *stability polynomial*. The integration scheme (5.96) is said to be *stable* when the roots of the stability equation (5.100) satisfy $|\lambda_i| \leqslant 1$. The scheme is said to be *strongly stable* if N of the roots satisfy $|\lambda_i| < 1$.

In view of our previous discussion, the idea behind the definition of strong stability becomes evident. The stability equation applies to the limiting case $h = 0$. However, if all but one of the roots of the stability equation satisfy $|\lambda_i| < 1$ then it is possible to find a small finite h such that these roots still lie inside the unit circle. Consequently the parasitic solutions are not troublesome.

From (5.94) it can be seen that all of the Adams–Bashforth and Adams–Moulton methods are strongly stable. However, Simpson's corrector formula is stable but not strongly stable.

5.1.6 Systems of equations and higher order equations

As noted in the introduction to this chapter, only initial-value problems are considered. The linear multi-step methods that we have derived for a single first-order differential equation can easily be applied to the system of N first-order equations

$$\mathbf{y}' = \mathbf{f}(x, \mathbf{y}) \qquad (5.101)$$

with initial conditions $\mathbf{y}(x_0) = \mathbf{y}_0$, where the prime denotes differentiation with respect to x, and $\mathbf{y}$, $\mathbf{f}$ are column vectors with N components. For example, to advance the solution from x_n to x_{n+1}, the third-order Adams–Bashforth predictor formula and third-order Adams–Moulton corrector formula can be applied to each of the equations

$$\frac{dy}{dx} = f_i(x, y_1, y_2, \ldots, y_N) \qquad (5.102)$$

for $i = 1(1)N$ in turn. This is demonstrated below with an example.

Higher order initial-value problems can also be solved using linear multi-step methods. However the higher order equation must first be converted to a system of first-order equations. This can be achieved by simple substitutions. For example, for the second-order initial-value problem

$$y'' = g(x, y, y'), \quad y(x_0) = y_0, \quad y'(x_0) = y_0',$$ (5.103)

introduce $u = y'$. Then (5.103) can be expressed as the system of two first-order equations

$$\begin{array}{ll} y' = u & y(x_0) = y_0 \\ u' = g(x, y, u) & u(x_0) = y_0'. \end{array}$$ (5.104)

This is a special case of the more general problem

$$\begin{array}{ll} y' = f(x, y, u) & y(x_0) = y_0 \\ u' = g(x, y, u) & u(x_0) = u_0 \end{array}$$ (5.105)

which, in turn, is (5.101) or (5.102) with $N = 2$ and the dependent variables relabelled.

Example 5.4 For the system of differential equations

$$\begin{array}{ll} u' = x + u - v^2 & u(0) = 1 \\ v' = x^2 - v + u^2 & v(0) = 1 \end{array}$$

use the Taylor series method to generate starting values for u and v at $x = 0.1, 0.2$. Then use the third-order Adams–Bashforth and Adams–Moulton formulae to estimate $u(0.3)$ and $v(0.3)$.

To make use of the Taylor expansion (5.40) we first generate the values of the derivatives at $x = x_0$, starting with u' and v' given in the problem. Thus by systematically differentiating and substituting numerical values we obtain

$$\begin{array}{ll} u' = x + u - v^2 & u'(0) = 0 \\ v' = x^2 - v + u^2 & v'(0) = 0 \\ u'' = 1 + u' - 2vv' & u''(0) = 1 \\ v'' = 2x - v' + 2uu' & v''(0) = 0 \\ u^{(3)} = u'' - 2v'^2 - 2vv'' & u^{(3)}(0) = 1 \\ v^{(3)} = 2 - v' + 2u'^2 + 2uu'' & v^{(3)}(0) = 4 \\ u^{(4)} = u^{(3)} - 6v'v'' - 2vv^{(3)} & u^{(4)}(0) = -7 \\ v^{(4)} = -v^{(3)} + 6u'u'' + 2uu^{(3)} & v^{(4)}(0) = -2. \end{array}$$

Using these numerical values in (5.40) with $h = 1/10$ we obtain

$$u(x_0 + r/10) = 1 + \frac{1}{2}\left(\frac{r}{10}\right)^2 + \frac{1}{6}\left(\frac{r}{10}\right)^3 - \frac{7}{24}\left(\frac{r}{10}\right)^4 + \cdots.$$

$$v(x_0 + r/10) = 1 + \frac{2}{3}\left(\frac{r}{10}\right)^3 - \frac{1}{12}\left(\frac{r}{10}\right)^4 + \cdots.$$

Noting that $x_0 = 0$ and setting $r = 1, 2$ in turn yields

$$u_1 = u(0.1) = 1.0051, \qquad u_2 = u(0.2) = 1.0209$$
$$v_1 = v(0.1) = 1.0007, \qquad v_2 = v(0.2) = 1.0052$$

to four decimal places.

Applying the third-order Adams–Bashforth formula (5.15) to each of the equations for u and v we obtain

$$u_{n+1}^{(0)} = u_n + \frac{h}{12}(23f_n - 16f_{n-1} + 5f_{n-2})$$

$$v_{n+1}^{(0)} = v_n + \frac{h}{12}(23g_n - 16g_{n-1} + 5g_{n-2})$$

where $f(x, u, v) = x + u - v^2$, $g(x, u, v) = x^2 - v + u^2$, $f_n = f(x_n, u_n, v_n)$ and $g_n = g(x_n, u_n, v_n)$. The index (0) distinguishes the initial predicted values from improved estimates that we calculate below. We will use these formulae to estimate $u_3 = u(0.3)$ and $v_3 = v(0.3)$. From the initial conditions and starting values we calculate

$$f_0 = 0.0000, \quad f_1 = 0.1037, \quad f_2 = 0.2105$$
$$g_0 = 0.0000, \quad g_1 = 0.0195, \quad g_2 = 0.0770$$

and using the above formulae with $n = 2$ we obtain the predicted values

$$u_3^{(0)} = 1.0474, \quad v_3^{(0)} = 1.0174.$$

To improve on these predicted values we apply the third-order Adams–Moulton formula (5.32) (also see (5.38)) to the differential equations for u and v, giving

$$u_{n+1}^{(k+1)} = u_n + \frac{h}{12}\{5f(x_{n+1}, u_{n+1}^{(k)}, v_{n+1}^{(k)}) + 8f_n - f_{n-1}\}$$

$$v_{n+1}^{(k+1)} = v_n + \frac{h}{12}\{5g(x_{n+1}, u_{n+1}^{(k+1)}, v_{n+1}^{(k)}) + 8g_n - g_{n-1}\},$$

where the indices in brackets indicate the iteration. From the predicted values for $u_3^{(0)}$, $v_3^{(0)}$ we will use formulae to generate the sequences $u_3^{(0)}, u_3^{(1)}, u_3^{(2)}, \ldots$ and $v_3^{(0)}, v_3^{(1)}, v_3^{(2)}, \ldots$ until convergence. Thus, as for the predictor formulae, we set $n = 2$. Then we set $k = 0$ to calculate $u_3^{(1)}, v_3^{(1)}$. We have

$$f(x_3, u_3^{(0)}, v_3^{(0)}) = f(0.3, 1.0474, 1.0174) = 0.3123$$

and substituting for f_2, f_1, which are given above, we obtain $u_3^{(1)} = 1.0471$. Also

$$g(x_3, u_3^{(1)}, v_3^{(0)}) = g(0.3, 1.0471, 1.0174) = 0.1690$$

and substituting for g_1 and g_2 we obtain $v_3^{(1)} = 1.0172$. Thus the first corrected values are

$$u_3^{(1)} = 1.0471, \quad v_3^{(1)} = 1.0172.$$

Further iteration with the corrector formula is not necessary since these values agree with the predictor values $u_3^0, v_3^{(0)}$ when rounded to three decimal places, i.e. to three decimal places we obtain

$$u_3 = 1.047, \quad v_3 = 1.017.$$

We can estimate the truncation error using a formula similar to (5.69). Equation (5.69) was derived for the fourth-order Adams–Bashforth/Adams–Moulton schemes. The similar equation for the third-order Adams–Bashforth/Adams–Moulton schemes (Exercise 15) is

$$y_{n+1} - w_{n+1} \simeq -\tfrac{1}{10}(w_{n+1} - w^{(0)}_{n+1}).$$

Applying this formula we find that estimates of the truncation are

$$-\tfrac{1}{10}(u_3 - u^{(0)}_3) \simeq 0$$
$$-\tfrac{1}{10}(v_3 - v^{(0)}_3) \simeq 0,$$

where $u^{(0)}_3$ and $v^{(0)}_3$ have been rounded to three decimal places before estimating the truncation error. This does not mean that there is no truncation error; it implies that the truncation error does not affect the third decimal places. Formulae such as (5.69) and the one used above will always yield zero estimates of the truncation error when the predictor formulae yield the answer to the required accuracy.

5.2 SINGLE-STEP METHODS

5.2.1 First-order equations

When multi-step methods are used to solve the initial-value problem

$$\frac{dy}{dx} = f(x, y), \quad y(x_0) = y_0, \tag{5.106}$$

the advancement of the solution from x_n to x_{n+1} is via formulae that involve estimates of the dependent variable at some of the grid points $x = x_{n-1}, x_{n-2}, \ldots$ as well as $x = x_n$. With the single-step methods that we describe here for solving the initial-value problem (5.106), the advancement of the solution from x_n to x_{n+1} is via formulae that only involve the estimate of the dependent variable at the previous grid point $x = x_n$, i.e. the entire calculation is based on the interval $[x_n, x_{n+1}]$, and the estimates $w_{n-1}, w_{n-2}, \ldots$ outside this interval and not directly involved in the calculation.

The methods that we describe fall into the class that is generally termed Runge–Kutta methods. Runge–Kutta formulae take the general form

$$w_{n+1} = w_n + h \sum_{i=0}^{p} \alpha_i f(x_n + \beta_i h, y_n + \gamma_i h), \tag{5.107}$$

where $\beta_0 = \gamma_0 = 0$, and the α's, β's and γ's are constants determined such that (5.107), when expanded as a power series in h, agrees with the Taylor expansion

$$y_{n+1} = y(x_n + h) = y_n + hy'_n + \frac{h^2}{2!}y''_n + \frac{h^3}{3!}y^{(3)}_n + \cdots \tag{5.108}$$

to as many terms as possible. The γ's in (5.107) are usually expressed as a linear combination of previously calculated values of the function.

The algebra involved in expanding (5.107) as a power series in h and comparing with (5.108) is extremely tedious, even for small values of p. We demonstrate the technique for the case $p = 1$. Thus we seek a formula of the form

$$w_{n+1} = w_n + \alpha_0 k_0 + \alpha_1 k_1 \tag{5.109}$$

where

$$k_0 = hf(x_n, y_n)$$
$$k_1 = hf(x_n + \beta h, y_n + \gamma k_0).$$

When comparing these formulae with (5.107) it should be noted that, to simplify the derivation, we have set $\gamma_1 = \gamma f(x_n, y_n)$ and dropped the suffix from β_1. The constants α_0, α_1, β, γ will be determined such that

$$y_{n+1} \simeq y_n + \alpha_0 k_0 + \alpha_1 k_1 \tag{5.110}$$

agrees with the Taylor expansion (5.108) to as high an order as possible. Thus, using a double Taylor expansion for k_1:

$$\frac{k_1}{h} = f + (\beta h f_x + \gamma k_0 f_y) + \tfrac{1}{2}(\beta^2 h^2 f_{xx} + 2\beta\gamma h k_0 f_{xy} + \gamma^2 k_0^2 f_{yy}) + O(h^3),$$

where all functions and derivatives are evaluated at (x_n, y_n). Substituting $k_0 = hf$ we obtain

$$k_1 = hf + h^2(\beta f_x + \gamma f f_y) + \frac{h^3}{2}(\beta^2 f_{xx} + 2\beta\gamma f f_{xy} + \gamma^2 f^2 f_{yy}) + O(h^4)$$

and substituting k_0 and k_1 into (5.110) gives

$$y_{n+1} \simeq y_n + h(\alpha_0 + \alpha_1)f + h^2\alpha_1(\beta f_x + \gamma f f_y)$$

$$+ \frac{h^3}{2}\alpha_1(\beta^2 f_{xx} + 2\beta\gamma f f_{xy} + \gamma^2 f^2 f_{yy}) + O(h^4). \tag{5.111}$$

Before comparing this with (5.108) we express (5.108) in terms of f. Thus we substitute

$$y' = f(x, y)$$
$$y'' = f_x + y' f_y = f_x + f f_y$$
$$y^{(3)} = f_{xx} + f_{xy}y' + f_x f_y + f_y y' f_y + f f_{xy} + f f_{yy}y'$$
$$= f_{xx} + 2f f_{xy} + f^2 f_{yy} + f_y(f_x + f f_y)$$

into (5.108) and obtain

$$y_{n+1} = y_n + hf + \frac{h^2}{2}(f_x + f f_y)$$

$$+ \frac{h^3}{6}(f_{xx} + 2f f_{xy} + f^2 f_{yy} + f_y(f_x + f f_y)) + O(h^4). \tag{5.112}$$

Comparing the coefficients of hf, $h^2 f_x$, $h^2 f f_y$ in (5.111) and (5.112) gives

$$\alpha_0 + \alpha_1 = 1, \quad \beta\alpha_1 = 1/2, \quad \gamma\alpha_1 = 1/2 \tag{5.113}$$

and provided these equations are satisfied, (5.109) is accurate to second order. We have three equations in four unknowns. We can therefore take one of the constants $\alpha_0, \alpha_1, \beta, \gamma$ as a free parameter and express the other constants in terms of this parameter. Setting $\alpha_1 = c$ we have

$$\alpha_0 = 1 - c, \quad \alpha_1 = c, \quad \beta = \frac{1}{2c}, \quad \gamma = \frac{1}{2c}. \tag{5.114}$$

and for any non-zero value of c (5.109) will be second order. Since c is a free parameter we check whether there is any value of c that will make (5.109) third order. Substituting for α_1, β, γ, the third-order term in (5.111) takes the form

$$\frac{h^3}{8c}(f_{xx} + 2ff_{xy} + f^2 f_{yy}) \tag{5.115}$$

and there is no value of c that will make this agree with the third-order term in (5.112). Therefore (5.109) cannot be better than second order.

Since c is a free parameter there are an infinite number of second-order Runge–Kutta methods. However, to obtain one of the most frequently used second-order Runge–Kutta methods we set $c = 1/2$ in (5.114). Then (5.109) takes the form

$$\begin{aligned}
w_{n+1} &= w_n + \tfrac{1}{2}(k_0 + k_1) \\
k_0 &= hf(x_n, y_n) \\
k_1 &= hf(x_n + h, y_n + k_0).
\end{aligned} \tag{5.116}$$

By comparing (5.115) with (5.112) we find that the truncation error associated with (5.116) is

$$\varepsilon_n = -\frac{h^3}{12}(y_n^{(3)} - 3f(x_n, y_n)y_n''). \tag{5.117}$$

This second-order method is sometimes called Heun's method.

Higher order methods can be derived in a similar manner but, as mentioned previously, the algebra is laborious. As for the second-order case there are an infinite number of Runge–Kutta methods for each order. One of the most commonly used schemes is the fourth-order Runge–Kutta method

$$\begin{aligned}
w_{n+1} &= w_n + \tfrac{1}{6}(k_0 + 2k_1 + 2k_2 + k_3) \\
k_0 &= hf(x_n, y_n) \\
k_1 &= hf(x_n + h/2, y_n + k_0/2) \\
k_2 &= hf(x_n + h/2, y_n + k_1/2) \\
k_3 &= hf(x_n + h, y_n + k_2).
\end{aligned} \tag{5.118}$$

Thus, for this scheme, the truncation error is of order h^5.

Algorithm 5.2. Runge–Kutta method (fourth order)

```
read(x0, y0, h, n);
x1 := x0 + h;
FOR i := 1 TO n DO
```

```
BEGIN
k0:= h* f(x0,y0);
k1:= h* f(x0 + h/2, y0 + k0/2);
k2:= h* f(x0 + h/2, y0 + k1/2);
k3:= h* f(x1, y0 + k2);
y1:= y0 + (k0 + 2*k1 + 2*k2 + k3)/6;
write('At', x1, 'y has the value', y1);
x0:= x1; x1:= x0 + h; y0:= y1
END.
```

Whereas some multi-step methods can exhibit numerical instability, this problem does not arise with one-step methods, such as those of the Runge–Kutta type, provided h is sufficiently small. The stability equation takes the form

$$\lambda - 1 = 0 \tag{5.119}$$

and consequently there are no parasitic solutions (see section 5.1.5(iii)).

Both equations (5.116) and (5.118) can be used to advance the solution, one step at a time, from an initial condition $y(x_0) = y_0$. Thus to calculate $w_1 \simeq y_1 = y(x_1) = y(x_0 + h)$, either scheme can be used with $n = 0$ (obviously the fourth-order scheme is more accurate). Similarly $w_2 \simeq y(x_0 + 2h)$, $w_3 \simeq y(x_0 + 3h)$,... can be calculated, in turn, be setting $n = 1, 2,...$ respectively in either scheme. An application of the fourth-order Runge–Kutta scheme (5.118) is given in Example 5.5 for a system of first-order equations.

5.2.2 Systems of equations and higher order equations

Runge–Kutta methods can easily be applied to the system of equations (5.101). For example, the fourth-order scheme (5.118) can be applied to each of the N equations in (5.102). As an example we consider the system of two first-order equations

$$\frac{dy}{dx} = f(x, y, u), \quad \frac{du}{dx} = g(x, y, u). \tag{5.120}$$

Since we cannot use w to represent numerical estimates of both y and u we use the corresponding capital letters instead. Thus the fourth-order Runge–Kutta scheme (5.118) applied to (5.120) is

$$Y_{n+1} = Y_n + \tfrac{1}{6}(k_0 + 2k_1 + 2k_2 + k_3)$$

$$U_{n+1} = U_n + \tfrac{1}{6}(m_0 + 2m_1 + 2m_2 + m_3) \tag{5.121}$$

$$k_0 = hf(x_n, Y_n, U_n)$$

$$m_0 = hg(x_n, Y_n, U_n)$$

$$k_1 = hf(x_n + h/2, Y_n + k_0/2, U_n + m_0/2)$$

$$m_1 = hg(x_n + h/2, Y_n + k_0/2, U_n + m_0/2)$$

$$k_2 = hf(x_n + h/2, Y_n + k_1/2, U_n + m_1/2)$$

$$m_2 = hg(x_n + h/2, Y_n + k_1/2, U_n + m_1/2)$$
$$k_3 = hf(x_n + h, Y_n + k_2, U_n + m_2)$$
$$m_3 = hg(x_n + h, Y_n + k_2, U_n + m_2)$$

Note that the ks and the ms must be calculated in the order given before advancing the solution for both y and u to $x = x_{n+1}$. In particular, it is not normally possible to calculate the ks and advance the solution for y before calculating the ms.

Higher-order equations must be reduced to a system of first-order equations before applying Runge–Kutta methods. This was discussed in section 5.16 for multi-step methods.

Example 5.5 Use a fourth-order Runge–Kutta method to calculate $u(0.1), v(0.1)$ for the initial value problem

$$u' = x + u - v^2 \quad u(0) = 1$$
$$v' = x^2 - v + u^2 \quad v(0) = 1$$

This is a direct application of the equations (5.121). For this problem we have

$$u' = f(x, u, v) = x + u - v^2$$
$$v' = g(x, u, v) = x^2 - v + u^2$$

and calculating the ks and ms in turn leads to

$$k_0 = hf(x_0, u_0, v_0)$$
$$= 0.1 f(0, 1, 1) = 0.00000$$
$$m_0 = 0.1g(0, 1, 1) = 0.00000$$

$$k_1 = hf(x_0 + h/2, u_0 + k_0/2, v_0 + m_0/2)$$
$$= 0.1 f(0.05, 1, 1) = 0.00500$$
$$m_1 = 0.1g(0.05, 1, 1) = 0.00025$$

$$k_2 = hf(x_0 + h/2, u_0 + k_1/2, v_0 + m_1/2)$$
$$= 0.1 f(0.05, 1.00250, 1.00013) = 0.00522$$
$$m_2 = 0.1g(0.05, 1.00250, 1.00013) = 0.00074$$

$$k_3 = hf(x_0 + h, u_0 + k_2, v_0 + m_2)$$
$$= 0.1 f(0.1, 1.00523, 1.00074) = 0.01037$$
$$m_3 = 0.1g(0.1, 1.00523, 1.00074) = 0.00197.$$

Finally (5.121) leads to

$$u(0.1) = u_1 = u_0 + \tfrac{1}{6}(k_0 + 2k_1 + 2k_2 + k_3)$$
$$= 1.00514$$
$$v(0.1) = v_1 = v_0 + \tfrac{1}{6}(m_0 + 2m_1 + 2m_2 + m_3)$$
$$= 1.00066$$

and to four decimal places we obtain

$$u(0.1) = 1.0051, \quad v(0.1) = 1.0007.$$

EXERCISES 5

1. From equation (5.8), derive equation (5.12) up to third-order differences, i.e. evaluate the coefficients (5.11) for $k = 1, 2, 3$.

2. From equation (5.12) derive the integration schemes (5.14)–(5.17).

3. For the integration scheme (5.16), what is the degree of the polynomial that has been used to approximate the integrand? Which points does this approximation pass through?

4. Use the definition of the backward difference operator (section 3.1) to derive equation (5.21).

5. From equation (5.26), derive equation (5.30) up to third-order differences, i.e. evaluate the coefficients (5.29) for $k = 1, 2, 3$.

6. From equation (5.30) derive the integration schemes (5.31)–(5.34).

7. Use the Taylor series method to estimate $y(\pm 0.1)$, $y(\pm 0.2)$ for each of the following initial-value problems:

   (i) $y' - y = 1 - 2x$, $y(0) = 2$
   (ii) $xy' - y = x^3$, $y(0) = 0$
   (iii) $y' - 2xy = 1 - 2x^2$, $y(0) = 1$
   (iv) $x^2 y' = xy + x^3 + xy^2$, $y(0) = 0$.

8. Repeat Exercise 7, (i)–(iv), using Picard's method.

9. Use either the Taylor series method or Picard's method to estimate $y(0.1)$, $y(0.2)$ and $y(0.3)$ for the initial-value problem 7(iii). Use the iterative improvement formulae (5.49)–(5.51) to check or improve these estimates.

10. By integrating Stirling's interpolation formulae, derive each of the following formulae to second differences:

$$y_{-2} = y_0 + h[-2 + 2\mu\delta - \tfrac{4}{3}\delta^2 + \tfrac{1}{3}\mu\delta^3 - \tfrac{7}{45}\delta^4 + \cdots]f_0$$
$$y_{-1} = y_0 + h[-1 + \tfrac{1}{2}\mu\delta - \tfrac{1}{6}\delta^2 - \tfrac{1}{24}\mu\delta^3 + \tfrac{1}{180}\delta^4 + \cdots]f_0$$
$$y_1 = y_0 + h[1 + \tfrac{1}{2}\mu\delta + \tfrac{1}{6}\delta^2 - \tfrac{1}{24}\mu\delta^3 - \tfrac{1}{180}\delta^4 + \cdots]f_0$$
$$y_2 = y_0 + h[2 + 2\mu\delta + \tfrac{4}{3}\delta^2 + \tfrac{1}{3}\mu\delta^3 + \tfrac{7}{45}\delta^4 + \cdots]f_0.$$

11. Use the formulae in Exercise 10 to check or improve the starting values for the initial-value problems 7(i)–7(iv).

12. With a step length of 0.1 use the Adams–Bashforth and Adams–Moulton formulae to advance the solutions for 7(i)–7(iv) to $x = 0.5$.

13. Derive Milne's predictor formulae, with error term,

$$w_{n+1} = w_{n-3} + \frac{4h}{3}(2f_n - f_{n-1} + 2f_{n-2}), \quad \varepsilon_{n+1} = \tfrac{14}{45}h^5 y^{(5)}(\xi)$$

and Simpson's corrector formula, with error term,

$$w_{n+1} = w_{n-1} + \frac{h}{3}(f_{n+1} + 4f_n + f_{n-1}), \quad \varepsilon_{n+1} = -\frac{h^5 y^{(5)}}{90}(\xi).$$

14. Repeat Exercise 12 using the Milne–Simpson predictor–corrector method instead of the Adams–Bashforth and Adams–Moulton formulae.

15. For the third-order Adams–Bashforth/Adams–Moulton formulae

$$w_{n+1}^{(0)} = w_n + \frac{h}{12}(23f_n - 16f_{n-1} + 5f_{n-2}), \quad \varepsilon_{n+1} = \tfrac{3}{8}h^4 y^{(4)}(\alpha_n)$$

$$w_{n+1} = w_n + \frac{h}{12}(5f_{n+1} + 8f_n - f_{n-1}), \quad \varepsilon_{n+1} = -\frac{h^4 y^{(4)}}{24}(\beta_n)$$

derive the equation

$$y_{n+1} - w_{n+1} = -\tfrac{1}{10}(w_{n+1} - w_{n+1}^{(0)})$$

for estimating the truncation error.

16. Show that, for the third-order Adams–Moulton corrector formula to converge, the step length h must be less than $12/(5f_y(x_n, y_n))$.

17. Show that, when Simpson's corrector formula is applied to the initial-value problem (5.79), the parasitic part of the solution is

$$(-1)^n \left\{ \frac{(ch)^5}{360} y_0 + \frac{\varepsilon}{2} \right\} e^{-cnh/3},$$

provided ch is small and where ε is the error inherent in y_1.

18. For the initial-value problems

(i) $y'' = -y, \ y(0) = 0, \ y'(0) = 1$
(ii) $2y'' = 3xy' - 9y + 9, \ y(0) = 1, \ y'(0) = -2$

estimate y for $x = 0.1(0.1)0.4$ using starting methods and predictor–corrector methods.

19. Use a fourth-order Runge–Kutta method to estimate y for $x = 0.1(0.1)0.4$ for the initial-value problems 7(i)–7(iv).

20. For the initial-value problems 18(i), (ii), use a fourth-order Runge–Kutta method to estimate y for $x = 0.1(0.1)0.4$.

6

Systems of Linear Equations

There is no difficulty in solving a simple system of linear equations, such as

$$x - 3y + 2z = 7$$
$$x + y - 3z = 6$$
$$\vdots$$
$$2x + 3y - z = 10,$$

using elementary elimination methods. However, problems frequently arise in which the system of equations is very large, e.g. 50 equations in 50 unknowns, or even larger. Although elementary elimination methods can be used to solve large linear systems they must be applied in a systematic manner. We describe this approach and some alternative methods of solution. The techniques available fall into one of two broad categories, namely direct methods or iterative methods. When direct methods are used, the solution can be obtained by carrying out a fixed number of calculations, and this number can be determined before starting the calculation. When iterative methods are used, the number of calculations depends on the accuracy required.

We will examine methods for solving the general system of n linear equations in n unknowns,

$$a_{11}x_1 + a_{12}x_2 + \cdots + a_{1n}x_n = b_1$$
$$a_{21}x_1 + a_{22}x_2 + \cdots + a_{2n}x_n = b_2 \qquad (6.1a)$$
$$\vdots$$
$$a_{n1}x_1 + a_{n2}x_2 + \cdots + a_{nn}x_n = b_n,$$

where the coefficients $a_{ij}(i = 1(1)n, j = 1(1)n)$ and $b_i(i = 1(1)n)$ are known constants for a specific problem and $x_i(i = 1(1)n)$ are the unknown quantities. It is often convenient to write the system (6.1a) in the matrix form

$$\mathbf{Ax = b} \qquad (6.1b)$$

where

$$A = \begin{bmatrix} a_{11} & a_{12} & \cdots & a_{1n} \\ a_{21} & a_{22} & \cdots & a_{2n} \\ \vdots & \vdots & & \vdots \\ a_{n1} & a_{n2} & \cdots & a_{nn} \end{bmatrix}, \quad b = \begin{bmatrix} b_1 \\ b_2 \\ \vdots \\ b_n \end{bmatrix}, \quad x = \begin{bmatrix} x_1 \\ x_2 \\ \vdots \\ x_n \end{bmatrix}. \tag{6.2}$$

Thus A is an $n \times n$ matrix, called the matrix of coefficients, b is a known column vector and x is the unknown column vector.

We will only consider systems of linear equations that have a unique solution. The theory of linear algebra shows us that the solution of (6.1) is unique when

(i) the system of linear equations (6.1b) is non-homogeneous, i.e. at least one element of the column vector b is non-zero (a system of linear equations is said to be homogeneous when every term in every equation contains an unknown quantity, and for (6.1b) this would require $b = 0$),

(ii) the matrix of coefficients is non-singular, i.e. det $A \neq 0$ (a square matrix is said to be singular when its determinant is zero).

Before describing the techniques that are often used for solving systems of linear equations we mention one that is rarely used. This is Cramer's rule, which states that

$$x_j = \frac{\det A_j}{\det A}, \quad j = 1(1)n, \tag{6.3}$$

where A_j is the $n \times n$ matrix with all columns, except the jth, identical to those of A. The jth column is identical to the column vector b. The calculation of large determinants is very time-consuming and consequently this method is not recommended for large values of n.

6.1 DIRECT METHODS

6.1.1 Triangular systems—forward and backward substitution

Direct methods usually involve reducing a given system of linear equations to triangular form. Consequently, before solving the general system of linear equations (6.1), we introduce the simple techniques of forward and backward substitution for solving triangular systems of linear equations.

Consider the upper triangular system

$$a_{11} + a_{12}x_2 + a_{13}x_3 + \cdots + a_{1n}x_n = b_1$$
$$a_{22}x_2 + a_{23}x_3 + \cdots + a_{2n}x_n = b_2$$
$$a_{33}x_3 + \cdots + a_{3n}x_n = b_3 \tag{6.4}$$
$$\vdots$$
$$a_{n-1\,n-1}x_{n-1} + a_{n-1\,n}x_n = b_{n-1}$$
$$a_{nn}x_n = b_n.$$

This system can be easily solved in reverse order. The nth equation gives $x_n = b_n/a_{nn}$. The $(n-1)$th equation only involves x_n and x_{n-1}. On substituting for x_n, x_{n-1} can be calculated. The values for x_n, x_{n-1} can be substituted into the $(n-2)$nd equation and

x_{n-2} can then be calculated. Proceeding in this way, we calculate $x_n, x_{n-1}, x_{n-2}, \ldots$ in that order from equation $n, n-1, n-2, \ldots$ respectively. Since this technique generates the solution in reverse order it is usually referred to as backward substitution.

The backward substitution method can be summarized in algorithm form. Suppose that $x_n, x_{n-1}, \ldots, x_{j+1}$ have been calculated from equations $n, n-1, \ldots, j+1$ respectively. The jth equation is

$$a_{jj}x_j + a_{jj+1}x_{j+1} + \cdots + a_{jn}x_n = b_j$$

and can be rearranged to give an expression for x_j:

$$x_j = \frac{1}{a_{jj}}\left\{b_j - \sum_{k=j+1}^{n} a_{jk}x_k\right\}.$$

This equation holds for $j = n-1(-1)1$. Thus the backward substitution algorithm is

$$x_n = \frac{b_n}{a_{nn}}, \quad x_j = \frac{1}{a_{jj}}\left\{b_j - \sum_{k=j+1}^{n} a_{jk}x_k\right\} \quad \text{for } j = n-1(-1)1. \tag{6.5}$$

The use of backward substitution is demonstrated in the following section on Gaussian elimination (see equations (6.11) and (6.12)).

The lower triangular system of linear equations

$$
\begin{aligned}
a_{11}x_1 &= b_1 \\
a_{21}x_1 + a_{22}x_2 &= b_2 \\
a_{31}x_1 + a_{32}x_2 + a_{33}x_3 &= b_3 \\
&\vdots \\
a_{n1}x_1 + a_{n2}x_2 + a_{n3}x_3 + \cdots + a_{nn}x_n &= b_n
\end{aligned}
\tag{6.6}
$$

can also be solved by simple substitution. Here the first equation is used to calculate x_1. This value is substituted into the second equation, which is used to calculate x_2. Proceeding in this way we calculate $x_1, x_2, x_3, \ldots, x_n$ from the 1st, 2nd, ..., nth equations respectively. This procedure is called forward substitution. The reader is left to check that the forward substitution algorithm is

$$x_1 = \frac{b_1}{a_{11}}, \quad x_j = \frac{1}{a_{jj}}\left\{b_j - \sum_{k=1}^{j-1} a_{jk}x_k\right\} \quad \text{for } j = 2(1)n. \tag{6.7}$$

6.1.2 Gaussian elimination

This is a systematic treatment of the elementary elimination process for solving linear simultaneous equations. The Gaussian elimination process reduces a given system of linear equations to an upper triangular system of equations which can be solved by backward substitution.

We demonstrate the method for the system of linear equations

$$
\begin{aligned}
2x_1 + x_2 + 4x_3 + 7x_4 &= 1 \\
-4x_1 + x_2 - 6x_3 - 13x_4 &= -1 \\
x_1 + 5x_2 + 7x_3 + 7x_4 &= 4 \\
-2x_1 + 5x_2 - 4x_3 - 8x_4 &= -5.
\end{aligned}
\tag{6.8}
$$

With the aim of producing an upper triangular system we eliminate x_1 from the second, third and fourth equations. To eliminate x_1 from the jth equation ($j = 2, 3, 4$) we scale the first equation so that the coefficient of x_1 takes the same value in the first and jth equations. Thus, since 2 is the coefficient of x_1 in the first equation, we multiply the first equation by $c_j/2$ for $j = 2, 3, 4$, in turn, where $c_2 = -4$, $c_3 = 1$, $c_4 = -2$ are the coefficients of x_1 in the second, third and fourth equations. The first equation is then subtracted from the jth equation for $j = 2, 3, 4$. This leads to the system of equations

$$\begin{aligned}
2x_1 + x_2 + 4x_3 + 7x_4 &= 1 \\
3x_2 + 2x_3 + x_4 &= 1 \\
4.5x_2 + 5x_3 + 3.5x_4 &= 3.5 \\
6x_2 \quad - x_4 &= -4.
\end{aligned} \tag{6.9}$$

We have used the first equation to eliminate x_1 from each of the other equations. Since this equation has played a key role in the elimination process it is given the special name of pivotal equation. Also, the coefficient of x_1 in the pivotal equation is an important number in the elimination process and is called the pivot. In particular the first equation was multiplied by $c_j/2$, i.e. divided by the pivot and multiplied by c_j. Division by the pivot can magnify or dampen round-off error. Below we discuss how to reduce round-off error by re-ordering the equations and changing the value of the pivot.

Again with the aim of producing an upper triangular system we now eliminate x_2 from the third and fourth equations in (6.9). The elimination procedure is similar to that described above. However, the first equation plays no part in this stage of the calculation. The second equation is the pivotal equation and 3, the coefficient of x_2 in the pivotal equation, is the pivot. Thus the multipliers for the pivotal equation are 4.5/3 and 6/3. With these multipliers, subtraction of the scaled pivotal equation from the third and fourth equations eliminates x_2 as required. This leads to the system of equations

$$\begin{aligned}
2x_1 + x_2 + 4x_3 + 7x_4 &= 1 \\
3x_2 + 2x_3 + x_4 &= 1 \\
2x_3 + 2x_4 &= 2 \\
-4x_3 - 3x_4 &= -6.
\end{aligned} \tag{6.10}$$

In passing we note that the common factor would normally be removed from the third equation for hand calculations. However we are describing a systematic procedure that could be programmed. Although a computer could be programmed to look for common factors in integer numbers, the Gaussian elimination procedure almost always generates decimal coefficients. Thus real arithmetic has to be used and integer coefficients are treated as real numbers. Hence the removal of common factors would be pointless and time-consuming.

Finally, to obtain an upper triangular form we eliminate x_3 from the fourth equation in (6.10). The first and second equations in (6.10) play no part in this stage of the calculation. The third equation is the pivotal equation and 2, the coefficient of x_3 in the pivotal equation, is the pivot. Thus multiplying the pivotal equation by $-4/2$ and subtracting from the fourth equation leads to the upper triangular system of equations

$$
\begin{aligned}
2x_1 + x_2 + 4x_3 + 7x_4 &= 1 \\
3x_2 + 2x_3 + x_4 &= 1 \\
2x_3 + 2x_4 &= 2 \\
x_4 &= -2.
\end{aligned}
\tag{6.11}
$$

This system of equations can be solved immediately by back substitution:

$$
\begin{aligned}
x_4 &= -2 \\
x_3 &= (2 - 2x_4)/2 = 3 \\
x_2 &= (1 - x_4 - 2x_3)/3 = -1 \\
x_1 &= (1 - 7x_4 - 4x_3 - x_2)/2 = 2,
\end{aligned}
\tag{6.12}
$$

i.e. the solution is

$$
x_1 = 2, \quad x_2 = -1, \quad x_3 = 3, \quad x_4 = -2.
\tag{6.13}
$$

Pivoting

Although one would expect direct methods to yield the exact solution of a system of linear equations, round-off errors can cause considerable inaccuracy. This is easily demonstrated with the system of two simultaneous equations

$$
\begin{aligned}
0.0001x_1 + 2.1517x_2 &= 2.1530 \\
0.1131x_1 - 2.5333x_2 &= -1.0630.
\end{aligned}
$$

It can be checked by substitution that the exact solution of this system of equations is $x_1 = 13, x_2 = 1$. We will now solve this system of equations using Gaussian elimination. Throughout the calculation we will work strictly to four significant figures. Thus we multiply the first equation by $0.1131/0.0001 = 1131$ and subtract the resulting equation from the second equation. This eliminates x_1 from the second equation and gives

$$
\begin{aligned}
x_2(-2.533 - (1131)(2.152)) \\
= -1.063 - (2.153)(1131) \\
x_2(-2437) = -2436.
\end{aligned}
$$

Thus $x_2 = 0.9996$ and substituting back into the first equation

$$
x_1 = \frac{2.153 - (2.152)(0.9996)}{0.0001}
$$

$$
= \frac{2.153 - 2.151}{0.0001}
$$

$$
= 20.
$$

This error in the value of x_1 arises because of round-off error in the calculations. Round-off can give rise to errors of considerably larger magnitude, particularly in larger systems of equations.

The effect of round-off error can be reduced by rearranging the equations at each stage of the elimination process. The rearrangement is such that the pivot is as large as possible. In the Gaussian elimination procedure, every coefficient in the pivotal equation is divided by the pivot. This division by the pivot will enhance or dampen any round-off error in the coefficients, depending on the size of the pivot. Clearly large pivots will reduce the effect of round-off error. Thus at the stage of the elimination process when x_j is to be eliminated, the equation in which x_j has the coefficient of largest magnitude is chosen as the pivotal equation and moved into the pivotal position. Obviously only those equations that do not contain $x_1, \ldots, x_{j-1}$ are scanned when searching for the largest coefficient of x_j. This procedure is called partial pivoting. As an example, the second equation in (6.8) would be chosen as the pivotal equation to eliminate x_1 from the remaining three equations. Consequently the first and second equations would swop positions.

Partial pivoting involves searching for the largest coefficient of an unknown quantity amongst a system of equations. This procedure is not well-defined since any equation can be multiplied by an arbitrary factor without altering the exact solution of the system. This can be overcome by normalizing each equation that is involved at each stage of the elimination procedure before carrying out partial pivoting. The normaliz-ation is such that the largest coefficient in all of these equations is unity. This is referred to as scaled partial pivoting and is demonstrated in the following example.

Example 6.1 Solve the system of equations (6.8) using scaled partial pivoting.

First we normalize each of the equations in (6.8) so that the magnitude of the largest coefficient in each equation is unity. Thus we divide each of the four equations by 7, $-13, 7$ and -8 respectively. Working to four decimal places this gives

$$0.2857x_1 + 0.1429x_2 + 0.5714x_3 + 1.0000x_4 = 0.1429$$
$$0.3077x_1 - 0.0769x_2 + 0.4615x_3 + 1.0000x_4 = 0.0769$$
$$0.1429x_1 + 0.7143x_2 + 1.0000x_3 + 1.0000x_4 = 0.5714$$
$$0.2500x_1 - 0.6250x_2 + 0.5000x_3 + 1.0000x_4 = 0.6250.$$

The coefficient of largest magnitude for x_1 occurs in the second equation. Thus the first two equations are interchanged:

$$0.3077x_1 - 0.0769x_2 + 0.4615x_3 + 1.0000x_4 = 0.0769$$
$$0.2857x_1 + 0.1429x_2 + 0.5714x_3 + 1.0000x_4 = 0.1429$$
$$0.1429x_1 + 0.7143x_2 + 1.0000x_3 + 1.0000x_4 = 0.5714$$
$$0.2500x_1 - 0.6250x_2 + 0.5000x_3 + 1.0000x_4 = 0.6250.$$

Multiples of the pivotal equation are now subtracted from the other three equations to eliminate x_1:

$$0.3077x_1 - 0.0769x_2 + 0.4615x_3 + 1.0000x_4 = 0.0769$$
$$0.2143x_2 + 0.1429x_3 + 0.0715x_4 = 0.0715$$
$$0.7500x_2 + 0.7857x_3 + 0.5356x_4 = 0.5357$$
$$-0.5625x_2 + 0.1250x_3 + 0.1875x_4 = 0.5625.$$

At this stage we normalize the last three equations before pivoting and eliminating x_2 from two of these equations. Thus the second, third and fourth equations are divided by

0.2143, 0.7857 and -0.5625 respectively, leading to

$$0.3077x_1 - 0.0769x_2 + 0.4615x_3 + 1.0000x_4 = \quad 0.0769$$
$$1.0000x_2 + 0.6668x_3 + 0.3336x_4 = \quad 0.3336$$
$$0.9546x_2 + 1.0000x_3 + 0.6817x_4 = \quad 0.6818$$
$$1.0000x_2 - 0.2222x_3 - 0.3333x_4 = -1.0000.$$

The second or fourth equations can be used as the pivotal equation. We use the second equation to eliminate x_2 from the third and fourth equations, giving

$$0.3077x_1 - 0.0769x_2 + 0.4615x_3 + 1.0000x_4 = \quad 0.0769$$
$$1.0000x_2 + 0.6668x_3 + 0.3336x_4 = \quad 0.3336$$
$$0.3635x_3 + 0.3632x_4 = \quad 0.3633$$
$$-0.8890x_3 - 0.6669x_4 = -1.3336.$$

We now normalize the third and fourth equations by dividing them by 0.3635 and -0.8890 respectively, leading to

$$0.3077x_1 - 0.0769x_2 + 0.4615x_3 + 1.0000x_4 = 0.0769$$
$$1.0000x_2 + 0.6668x_3 + 0.3336x_4 = 0.3336$$
$$1.0000x_3 + 0.9992x_4 = 0.9994$$
$$1.0000x_3 + 0.7502x_4 = 1.5001.$$

Examining the coefficients in the third and fourth equations we see that no rearrangement is necessary at this stage. Thus we use the third equation to eliminate x_3 from the fourth equation. This leads to the triangular system

$$0.3077x_1 - 0.0769x_2 + 0.4615x_3 + 1.0000x_4 = 0.0769$$
$$1.0000x_2 + 0.6668x_3 + 0.3336x_4 = 0.3336$$
$$1.0000x_3 + 0.9992x_4 = 0.9994$$
$$-0.2490x_4 = 0.5007.$$

Sovling by back substitution leads to

$$x_4 = -2.0108, \quad x_3 = 3.0086, \quad x_2 = -1.0017, \quad x_1 = 2.0221. \qquad (6.14)$$

The purpose of Example 6.1 was to demonstrate how to use scaled partial pivoting. It does not demonstrate its effectiveness. When we compare the two solutions (6.13) and (6.14) it appears that the use of scaled partial pivoting leads to less accurate solutions rather than more accurate solutions. However, this is not a fair comparison for assessing scaled partial pivoting, since the solution (6.13) was obtained entirely by integer arithmetic. Since decimals were not involved, there was no round-off and consequently exact solutions were obtained. It is only with simple systems of equations that the solution can be obtained via integer arithmetic.

Algorithm 6.1. Gaussian elimination with partial pivoting

The non-singular matrix A is augmented by making b its $(n + 1)$th column.

```
read(n);
FOR i:= 1 TO n DO
  FOR j:= 1 TO n + 1 DO read(A[i, j]);
FOR i:= 1 TO n DO
  BEGIN
(*normalize each row*)
  FOR k:= i TO n DO
    BEGIN
    MaxCol:= i; MaxVal:= abs(A[k, i]);
    FOR j:= i + 1 TO n DO
      IF abs(A[k, j]) > MaxVal THEN
        BEGIN
        MaxCol:= j; MaxVal:= abs(A[k, j])
        END;
    Scale:= 1/A[k, MaxCol];
    FOR j:= i TO n + 1 DO A[k, j]:= Scale*A[k, j]
    END;
(*find pivotal row*)
  MaxRow:= i; MaxVal:= abs(A[i, i]);
  FOR k:= i + 1 TO n DO
    IF abs(A[k, i]) > MaxVal THEN
      BEGIN
      MaxRow:= k; MaxVal:= abs(A[k, i])
      END;
  If MaxRow < > i THEN
    FOR j:= i TO n + 1 DO
      BEGIN
      Temp:= A[i, j]; A[i, j]:= A[MaxRow, j];
      A[MaxRow, j]:= Temp
      END;
(*zero remainder of column*)
  InvAii:= 1/A[i, i];
  For k:= i + 1 TO n DO
    BEGIN
    Scale:= InvAii*A[k, i];
    FOR j:= i TO n + 1 DO A[k, j]:= A[k, j] − Scale*A[i, j]
    END
  END;
(*back substitution*)
x[n]:= A[n, n + 1]/A[n, n];
FOR i:= n − 1 DOWNTO 1 DO
  BEGIN
  xi:= A[i, n + 1];
  FOR j:= i + 1 TO n DO xi:= xi − A[i, j]*x[j];
  x[i]:= xi/A[i, i]
  END
write(x).
```

6.1.3 Triangular factorization

This is an efficient direct method for solving the system of equations (6.1). It is particularly efficient when several systems of equations with the same matrix of coefficients have to be solved. The method involves finding lower and upper triangular matrices

$$\mathbf{L}_{n \times n} = (l_{ij}), \quad \mathbf{U}_{n \times n} = (u_{ij}) \tag{6.15a}$$

where

$$l_{ij} = 0 \quad \text{for } i < j, \quad u_{ij} = 0 \quad \text{for } i > j \tag{6.15b}$$

such that

$$\mathbf{A} = \mathbf{LU}. \tag{6.16}$$

The form of the matrices $\mathbf{L}$ and $\mathbf{U}$, defined in (6.15), can be seen below in equations (6.18) and (6.19).

We will describe the procedure for calculating $\mathbf{L}$ and $\mathbf{U}$ later. At this stage we will assume that $\mathbf{L}$ and $\mathbf{U}$ are known and show that the solution to (6.1) can be obtained fairly easily by forward and backward substitution. Substituting (6.16) into (6.1b) gives

$$\mathbf{LUx} = \mathbf{b}. \tag{6.17}$$

Defining the column vector $\mathbf{z}$ by

$$\mathbf{z} = \mathbf{Ux}, \tag{6.18}$$

i.e.

$$\begin{bmatrix} u_{11} & u_{12} & u_{13} & \cdots & u_{1n} \\ 0 & u_{22} & u_{23} & \cdots & u_{2n} \\ 0 & 0 & u_{33} & \cdots & u_{3n} \\ & & \vdots & & \\ 0 & 0 & 0 & \cdots & u_{nn} \end{bmatrix} \begin{bmatrix} x_1 \\ x_2 \\ x_3 \\ \vdots \\ x_n \end{bmatrix} = \begin{bmatrix} z_1 \\ z_2 \\ z_3 \\ \vdots \\ z_n \end{bmatrix}.$$

we write (6.17) in the form

$$\mathbf{Lz} = \mathbf{b}, \tag{6.19a}$$

i.e.

$$\begin{bmatrix} l_{11} & 0 & & \cdots & 0 \\ l_{21} & l_{22} & 0 & \cdots & 0 \\ l_{31} & l_{32} & l_{33} & \cdots & 0 \\ & & \vdots & & \\ l_{n1} & l_{n2} & l_{n3} & \cdots & l_{nn} \end{bmatrix} \begin{bmatrix} z_1 \\ z_2 \\ z_3 \\ \vdots \\ z_n \end{bmatrix} = \begin{bmatrix} b_1 \\ b_2 \\ b_3 \\ \vdots \\ b_n \end{bmatrix}. \tag{6.19b}$$

Thus to solve the system of equations (6.1) we first calculate $\mathbf{z}$ by solving the system of equations (6.19). This solution can be obtained using the forward substitution algorithm (see (6.7))

$$z_1 = \frac{b_1}{l_{11}}, \quad z_j = \frac{1}{l_{jj}} \left\{ b_j - \sum_{k=1}^{j-1} l_{jk} z_k \right\} \quad \text{for } j = 2(1)n. \tag{6.20}$$

The required solution $\mathbf{x}$ can then be found by solving the upper triangular system of equations (6.18). This system of equations can be solved by the backward substitution algorithm (see 6.5))

$$x_n = \frac{z_n}{u_{nn}}, \quad x_j = \frac{1}{u_{jj}}\left\{z_j - \sum_{k=j+1}^{n} u_{jk}x_k\right\} \quad \text{for } j = n-1(-1)1. \tag{6.21}$$

We now discuss the calculation of the matrices L and U. The matrices L and U are not uniquely determined by the matrix equation (6.16). These two matrices together contain $n^2 + n$ unknown elements (this is most easily seen by sliding the two matrices on top of each other—the overlapping diagonals give the additional n unknown elements). Thus when comparing elements on the left- and right-hand sides of (6.16) we have n^2 equations and $n^2 + n$ unknowns. Consequently we require a further n conditions to uniquely determine the matrices L and U. There are three additional sets of n conditions that are commonly used. These are

$$\text{Doolittle's method:} \quad l_{ii} = 1, \quad i = 1(1)n \tag{6.22}$$

$$\text{Choleski's method:} \quad l_{ii} = u_{ii}, \quad i = 1(1)n \tag{6.23}$$

$$\text{Crout's method:} \quad u_{ii} = 1, \quad i = 1(1)n. \tag{6.24}$$

There is no major difference between the three methods or significant advantage of one method over the others. We will describe Doolittle's method. To summarize the problem, we wise to determine L and U such that $A = LU$, where

$$A = (a_{ij})_{n \times n}, \quad L = (l_{ij})_{n \times n}, \quad U = (u_{ij})_{n \times n}$$

and

$$L = \begin{bmatrix} 1 & 0 & 0 & \cdots & \\ l_{21} & 0 & 0 & \cdots & \\ l_{31} & l_{32} & 1 & \cdots & \\ & & \vdots & & \\ l_{n1} & l_{n2} & l_{n3} & \cdots & 1 \end{bmatrix}, \quad U = \begin{bmatrix} u_{11} & u_{12} & u_{13} & \cdots & u_{1n} \\ 0 & u_{22} & u_{23} & \cdots & u_{2n} \\ 0 & 0 & u_{33} & \cdots & u_{3n} \\ & \vdots & & & \vdots \\ 0 & 0 & 0 & \cdots & u_{nn} \end{bmatrix}. \tag{6.25}$$

By matrix multiplication we see that

$$a_{ij} = \sum_{k=1}^{n} l_{ik}u_{kj}$$

and since $l_{ik} = 0$ for $i < k$ and $u_{kj} = 0$ for $k > j$ this reduces to

$$a_{ij} = \sum_{k=1}^{\min(i,j)} l_{ik}u_{kj}$$

or

$$a_{ij} = l_{i1}u_{1j} + l_{i2}u_{2j} + \cdots + l_{ii}u_{ij} = \sum_{k=1}^{i} l_{ik}u_{kj} \quad \text{for } i \leqslant j \tag{6.26}$$

$$a_{ij} = l_{i1}u_{1j} + l_{i2}u_{2j} + \cdots + l_{ij}u_{jj} = \sum_{k=1}^{j} l_{ik}u_{kj} \quad \text{for } j < i. \tag{6.27}$$

Equations (6.26) and (6.27) constitute the n^2 equations that must be solved to determine the elements of L and U. Although these equations appear to be non-linear we find that the elements of L and U can be determined in a systematic manner.

First we consider the top row of **A**. Thus we set $i = 1$ in (6.26) and obtain

$$a_{1j} = l_{11}u_{1j} = u_{1j} \quad \text{for } j = 1(1)n,$$

noting that $l_{11} = 1$. Hence the top row of **U** is given by

$$u_{1j} = a_{1j} \quad \text{for } j = 1(1)n. \tag{6.28}$$

Now consider the first column of **A**. Thus we set $j = 1$ in (6.27) and obtain

$$a_{i1} = l_{i1}u_{11} \quad \text{for } i = 2(1)n.$$

Hence the first column of **L** is given by

$$l_{i1} = a_{i1}/u_{11} \quad \text{for } i = 2(1)n. \tag{6.29}$$

To calculate the other elements of **L** and **U** we rearrange equation (6.26) to give an expression for u_{ij} and we rearrange equation (6.27) to give an expression for l_{ij}. Noting that $l_{ii} = 1$ this gives

$$u_{ij} = a_{ij} - \sum_{k=1}^{i-1} l_{ik}u_{kj}$$

$$l_{ij} = \frac{1}{u_{jj}} \left\{ a_{ij} - \sum_{k=1}^{j-1} l_{ik}u_{kj} \right\}$$

or, writing out the summations in full,

$$u_{ij} = a_{ij} - (l_{i1}u_{1j} + l_{i2}u_{2j} + \cdots + l_{i1-1}u_{i-1j}) \tag{6.30}$$

$$l_{ij} = \frac{1}{u_{jj}} [a_{ij} - (l_{i1}u_{1j} + l_{i2}u_{2j} + \cdots + l_{ij-1}u_{j-1j})] \tag{6.31}$$

Thus to calculate u_{ij} we need to know

$$u_{1j}, u_{2j}, \ldots, u_{i-1j}$$

which lie in

$$\text{column } j, \text{ rows } 1, 2, \ldots, i-1$$

of **U**. These will be known if the previous rows, 1 to $i - 1$, have been calculated. We also need

$$l_{i1}, l_{i2}, \ldots, l_{ii-1}$$

which lie in

$$\text{row } i, \text{ column } 1, 2, \ldots, i-1$$

of **L**. These will be known if columns 1 to $i - 1$ of **L** have been calculated. In summary, to calculate the ith row of U we need to know the first $i - 1$ rows of **U** and the first $i - 1$ columns of **L**.

Referring to (6.31), to calculate l_{ij} we need to know the values of

$$l_{i1}, l_{i2}, \ldots, l_{ij-1}$$

which lie in

$$\text{row } i, \text{ columns } 1, 2, \ldots, j-1$$

of **L**. These will be known if the first $j - 1$ columns of **L** have been calculated. We also need to know

$$u_{1j}, u_{2j}, \ldots, u_{j-1\,j}, u_{jj}$$

which lie in

$$\text{column } j, \text{ rows } 1, 2, \ldots, j$$

of U. Note that the value of l_{ij} depends on u_{jj}. Consequently the entire jth row of U is usually calculated first and then the jth column of L is calculated (this situation is reversed when Crout's method is used).

Thus the procedure for calculating the elements of L and U is to calculate the ith row of U followed by the ith column of L for $i = 1(1)n$. The calculation is via equations (6.28)–(6.31) which we summarize:

(i) first row of U

$$u_{1j} = a_{1j} \quad \text{for } j = 1(1)n \tag{6.32a}$$

(ii) first column of L

$$l_{j1} = a_{j1}/u_{11} \quad \text{for } j = 2(1)n \tag{6.32b}$$

(iii) ith row of U

$$u_{ij} = a_{ij} - \sum_{k=1}^{i-1} l_{ik} u_{kj} \tag{6.32c}$$

(iv) ith column of L

$$l_{ji} = \frac{1}{u_{ii}} \left\{ a_{ji} - \sum_{k=1}^{i-1} l_{jk} u_{ki} \right\}. \tag{6.32d}$$

Equations (6.32c) and (6.32d) are applied for $i = 2(1)n$ and for each value of i, $j = i(1)n$ for the u-calculations and $j = i + 1(1)n$ for the l-calculations (i specifies the row of U and column of L that is being calculated and j specifies a particular element within each row and column).

Algorithm 6.2 implements this scheme for calculating the lower and upper triangular matrices via Doolittle's method. The algorithm can be extended to find the solution of the system of equations (6.1) by adding the forward and backward substitution routines (6.20) and (6.21). This is left as an exercise for the reader.

Algorithm 6.2. Triangular factorization

The matrix is $A[1 \ldots n, 1 \ldots n]$ and its factors are $L[1 \ldots n, 1 \ldots n]$ & $U[1 \ldots n, 1 \ldots n]$. No zero divisors are assumed.

```
read(n);
FOR i:= 1 TO n DO
   FOR j:= 1 TO n DO read(A[i, j]);
FOR i:= 1 TO n DO
   BEGIN
   Uii:= A[i, i];
```

```
FOR k:= 1 TO i − 1 DO  Uii:= Uii − L[i,k]*U[k,i];
U[i,i]:= Uii; L[i,i]:= 1;
FOR j:= i + 1 TO n DO
  BEGIN
  U[j,i]:= 0; L[i,j]:= 0;
  Uij:= A[i,j]; Lji:= A[j,i];
  FOR k:= 1 TO i − 1 DO
    BEGIN
    Uij:= Uij − L[i,k]*U[k,j];
    Lji:= Lji − L[j,k]*U[k,i]
    END
  U[i,j]:= Uij; L[j,i]:= Lji/Uii
  END
END;
FOR i:= 1 TO n DO
  FOR j:= 1 TO n DO write(L[i,j]);
FOR i:= 1 TO n DO
  FOR j:= 1 TO n DO write(U[i,j]);.
```

Example 6.2 Use triangular factorization to solve the system of equations

$$10x_1 - 3x_2 + 2x_3 + 5x_4 = 30$$
$$3x_1 + 3x_2 + x_3 - x_4 = 5$$
$$4x_1 + 3x_2 + 7x_3 + 2x_4 = 10$$
$$x_1 - x_2 + x_3 + 2x_4 = 6.$$

We use equations (6.32) to estimate the rows of U and columns of L in turn. To complete these upper and lower triangular matrices we simultaneously fill in the zero elements of the rows and columns and the unit elements on the leading diagonal of L.
    The matrix of coefficients is

$$A = \begin{bmatrix} 10 & -3 & 2 & 5 \\ 3 & 3 & 1 & -1 \\ 4 & 3 & 7 & 2 \\ 1 & -1 & 1 & 2 \end{bmatrix}$$

and from equation (6.32a) the first row of U is given by

$$U_1 = [10 \quad -3 \quad 2 \quad 5].$$

From (6.23b) we obtain the first column of L by dividing the first column of A by $u_{11} = 10$:

$$l_1 = [1 \quad 0.3 \quad 0.4 \quad 0.1]^T.$$

We now use (6.32c), (6.32d) alternatively with $i = 2, 3, 4$ to calculate the remaining rows and columns of L and U. Thus the elements in the second row of U are given by

$$u_{2j} = a_{2j} - l_{21}u_{1j}$$
$$= a_{2j} - 0.3u_{1j}$$

for $j = 2, 3, 4$. The elements u_{1j}, for $j = 2, 3, 4$, have already been calculated (the first row of **U**). Hence the second row of **U** is

$$\mathbf{u}_2 = [0 \quad 3.9 \quad 0.4 \quad -2.5].$$

Using (6.32d) the elements in the second column of **L** are given by

$$l_{j2} = (a_{j2} - l_{j1}u_{12})/u_{22}$$
$$= (a_{j2} + 3l_{j1})/3.9$$

for $j = 3, 4$. Thus the second column of **L** is

$$l_2 = [0 \quad 1 \quad 1.0769 \quad -0.1795]^{\mathrm{T}}.$$

Returning to (6.32c) the elements in the third row of **U** are given by

$$u_{3j} = a_{3j} - \sum_{k=1}^{2} l_{3k} u_{kj}$$

$$= a_{3j} - (0.4u_{1j} + 1.0769u_{2j})$$

for $j = 3, 4$ and consequently the third row is

$$\mathbf{u}_3 = [0 \quad 0 \quad 5.7692 \quad 2.6923].$$

Using (6.32d) the only element to be calculated in the third column of **L** is

$$l_{j3} = \left(a_{j3} - \sum_{k=1}^{2} l_{jk} u_{k3} \right) / u_{33}$$

$$= (a_{j3} - 2l_{j1} - 0.4l_{j2})/5.7692$$

for $j = 4$. Hence the third column of **L** is

$$l_3 = [0 \quad 0 \quad 1 \quad 0.1511].$$

Using (6.32c) the only element to calculate in the fourth row of **U** is

$$u_{4j} = a_{4j} - \sum_{k=1}^{3} l_{4k} u_{kj}$$

$$= a_{4j} - (0.1u_{1j} - 0.1795u_{2j} + 0.1511u_{3j})$$

with $j = 4$. Hence the fourth row of **U** is

$$\mathbf{u}_4 = [0 \quad 0 \quad 0 \quad 0.6444].$$

No calculation is needed for the fourth column of **L** which is

$$l_4 = [0 \quad 0 \quad 0 \quad 1].$$

Thus the matrices **L** and **U** are given by

$$\mathbf{L} = \begin{bmatrix} 1 & 0 & 0 & 0 \\ 0.3 & 1 & 0 & 0 \\ 0.4 & 1.0769 & 1 & 0 \\ 0.1 & -0.1795 & 0.1511 & 1 \end{bmatrix}$$

$$U = \begin{bmatrix} 10 & -3 & 2 & 5 \\ 0 & 3.0 & 0.4 & -2.5 \\ 0 & 0 & 5.7692 & 2.6923 \\ 0 & 0 & 0 & 0.6444 \end{bmatrix}.$$

Note that the calculation can be checked by forming the product LU which should give the original matrix of coefficients A.

For this problem, equation (6.19b) takes the form

$$\begin{bmatrix} 1 & 0 & 0 & 0 \\ 0.3 & 1 & 0 & 0 \\ 0.4 & 1.0769 & 1 & 0 \\ 0.1 & -1.1795 & 0.1511 & 1 \end{bmatrix} \begin{bmatrix} z_1 \\ z_2 \\ z_3 \\ z_4 \end{bmatrix} = \begin{bmatrix} 30 \\ 5 \\ 10 \\ 6 \end{bmatrix}$$

and, using the forward substitution algorithm (6.20) (note $l_{jj} = 1$ for all j), we calculate, in turn,

$$z_1 = b_1 = 30$$
$$z_2 = b_2 - l_{21}z_1$$
$$= 5 - 0.3 \times 30$$
$$= -4$$

$$z_3 = b_3 - (l_{31}z_1 + l_{32}z_2)$$
$$= 10 - (0.4 \times 30 + 1.0769 \times (-4))$$
$$= 2.3076$$

$$z_4 = b_4 - (l_{41}z_1 + l_{42}z_2 + l_{43}z_3)$$
$$= 6 - (0.1 \times 30 + (-0.1795) \times (-4) + 0.1511 \times 2.3076)$$
$$= 1.9333.$$

Hence the system of equations (6.18) takes the form

$$\begin{bmatrix} 10 & -3 & 2 & 5 \\ 0 & 3.9 & 0.4 & -2.5 \\ 0 & 0 & 5.7692 & 2.6923 \\ 0 & 0 & 0 & 0.6444 \end{bmatrix} \begin{bmatrix} x_1 \\ x_2 \\ x_3 \\ x_4 \end{bmatrix} = \begin{bmatrix} 30 \\ -4 \\ 2.3076 \\ 1.9333 \end{bmatrix}$$

and, using the backward substitution algorithm (6.21), we calculate, in turn,

$$x_4 = z_4/u_{44}$$
$$= 1.9333/0.64444$$
$$= 3.0002$$

$$x_3 = (z_3 - u_{34}x_4)/u_{33}$$
$$= (2.3076 - 2.6923 \times 3.0002)/5.7692$$
$$= -1.0001$$

$$x_2 = (z_2 - u_{23}x_3 - u_{24}x_4)/u_{22}$$
$$= (-4 - 0.4(-1.0001) + 2.5 \times 3.0002)/3.9$$
$$= 1.0001$$

$$x_1 = (z_1 - u_{12}x_2 - u_{13}x_3 - u_{14}x_4)u_{11}$$
$$= (30 + 3 \times 1.0001 - 2 \times (-1.0001) - 5 \times 3.0002)/10$$
$$= 2.0000.$$

Since we have carried out all intermediate working to four decimal places we round our final results to three decimal places to obtain

$$x_1 = 2.000, \quad x_2 = 1.000, \quad x_3 = -1.000, \quad x_4 = 3.000.$$

It can be checked by substitution that the original system of equations has the integer solutions

$$x_1 = 2, \quad x_2 = 1, \quad x_3 = -1, \quad x_4 = 3.$$

6.2 ITERATIVE TECHNIQUES

With non-linear equations, such as those encountered in Chapter 2, iterative techniques may be the only means of obtaining the solution to an equation or system of equations. With linear equations the solution can usually be obtained either by direct or by iterative methods. Direct methods may seem to be the obvious choice for obtaining the solution of a system of linear equations. However there are problems for which iterative techniques provide an equally efficient or more efficient means of obtaining the solution. For linear systems of small dimension, iterative techniques are seldom used, since the time required for sufficient accuracy exceeds that required by direct methods. However, for large linear systems with a sparse matrix of coefficients (a matrix with a large number of zero entries is said to be sparse) iterative techniques can be efficient in terms of computer storage and time requirements. Such systems of equations arise frequently in the numerical solution of boundary-value problems and partial differential equations.

We want to obtain an iterative scheme that will generate a sequence of estimates $x_i^{(k)}$, $i = 1(1)n$, with the intention that

$$x_i^{(k)} \to x_i, \quad i = 1(1)n \text{ as } k \to \infty.$$

This can be achieved by relating new estimates $x_i^{(k+1)}$, $i = 1(1)n$, to known estimates $x_i^{(k)}$. There are various ways of doing this for a given linear system of equations. Some of the commonly used methods are presented in sections 6.2.1, 6.2.2 and 6.2.3. The iterative technique described in section 6.2.5, namely iterative improvement, is a special method for detecting and removing round-off error.

6.2.1 Jacobi's method

This scheme can be obtained by simply rearranging the system of equations (6.1a) such that x_i is isolated on the left-hand side of the ith equation for $i = 1(1)n$,

$$x_i = \left\{ b_i - \sum_{j=1}^{i-1} a_{ij}x_j - \sum_{j=i+1}^{n} a_{ij}x_j \right\} / a_{ii}, \quad i = 1(1)n. \tag{6.34}$$

and indicating the iteration by attaching '$k+1$' to the isolated x and attaching 'k' to all other x's on the right-hand side,

$$x_i^{(k+1)} = \left\{ b_i - \sum_{j=1}^{i-1} a_{ij}x_j^{(k)} - \sum_{j=i+1}^{n} a_{ij}x_j^{(k)} \right\} / a_{ii}, \quad i = 1(1)n. \tag{6.34}$$

Thus, in full, the Jacobi iterative scheme is

$$x_1^{(k+1)} = (b_1 - a_{12}x_2^{(k)} - a_{13}x_3^{(k)} - \cdots - a_{1n}x_n^{(k)})/a_{11}$$
$$x_2^{(k+1)} = (b_2 - a_{21}x_1^{(k)} - a_{23}x_3^{(k)} - \cdots - a_{2n}x_n^{(k)})/a_{22}$$
$$\vdots \qquad \vdots \qquad\qquad \vdots$$
$$x_i^{(k+1)} = (b_i - a_{i1}x_1^{(k)} - a_{ii-1}x_{i-1}^{(k)} - a_{ii+1}x_{i+1}^{(k)} - a_{in}x_n^{(k)})/a_{ii} \tag{6.35}$$
$$\vdots \qquad \vdots \qquad\qquad \vdots$$
$$x_n^{(k+1)} = (b_n - a_{n1}x_1^{(k)} - a_{n2}x_2^{(k)} - \cdots - a_{nn-1}x_{n-1}^{(k)})/a_{nn}$$

and, from an initial guess $x^{(0)}$ to the solution vector x, this scheme will provide successive estimates $x^{(k+1)}$, $k = 0, 1, \ldots$, to the solution vector x, i.e. (6.35) with $k = 0$ gives $x^{(1)}$ in terms of $x^{(0)}$, with $k = 1$ gives $x^{(2)}$ in terms of $x^{(1)}$, and so on.

If the system of equations $Ax = b$ has arisen from a physical or engineering problem, for example, information relating to the specific problem may help when making an initial guess. Failing this, simple initial guesses such as

$$x^{(0)} = (0 \quad 0 \quad \cdots \quad 0)^T \tag{6.36}$$

or

$$x^{(0)} = (1 \quad 1 \quad \cdots \quad 1)^T \tag{6.37}$$

are commonly used. Equivalent to (6.36), the initial guess

$$x_i^{(0)} = b_i/a_{ii}, \quad i = 1(1)n, \tag{6.38}$$

can also be used. Note that (6.38) is the estimate that arises when (6.36) is substituted into (6.35).

6.2.2 The Gauss–Siedel method

This is a modification of Jacobi's method that makes use of the latest estimates of the components of the solution vector as soon as they are available. When calculating $x_i^{(k+1)}$, the ith component of the solution vector at stage $k + 1$ of the iteration, the estimates $x_1^{(k+1)}, x_2^{(k+1)}, \ldots, x_{i-1}^{(k+1)}$ have already been calculated and are used instead of the estimates $x_1^{(k)}, x_2^{(k)}, \ldots, x_{i-1}^{(k)}$ which are used in Jacobi's method. Both schemes have no alternative but to use the kth estimates of $x_{i+1}, x_{i+2}, \ldots, x_n$ when calculating $x_i^{(k+1)}$. Thus the Gauss–Siedel method is fully described by the equation

$$x_i^{(k+1)} = \left\{ b_i - \sum_{j=1}^{i-1} a_{ij}x_j^{(k+1)} - \sum_{j=i+1}^{n} a_{ij}x_j^{(k)} \right\} \bigg/ a_{ii}, \quad i = 1(1)n, \tag{6.39}$$

which should be compared with Jacobi's method described in (6.34). Writing out the Gauss–Siedel iterative scheme in full, we have

$$x_1^{(k+1)} = (b_1 - a_{12}x_2^{(k)} - a_{13}x_3^{(k)} - \cdots - a_{1n}x_n^{(k)})/a_{11}$$
$$x_2^{(k+1)} = (b_2 - a_{21}x_1^{(k+1)} - a_{23}x_3^{(k)} - \cdots - a_{2n}x_n^{(k)})/a_{22}$$
$$x_i^{(k+1)} = (b_i - a_{ii}x_1^{(k+1)} - a_{ii-1}x_{i-1}^{(k+1)} - a_{ii+1}x_{i+1}^{(k)} - \cdots - a_{in}x_n^{(k)})/a_{ii} \tag{6.40}$$
$$x_n^{(k+1)} = (b_i - a_{n1}x_1^{(k+1)} - a_{n2}x_2^{(k+1)} - \cdots - a_{nn-1}x_n^{(k+1)})/a_{nn},$$

which should be compared with the Jacobi scheme in (6.35).

Example 6.3 Use the Gauss–Siedel method to solve the system of equations

$$-19x_1 + x_2 + 5x_3 + 3x_4 = -34$$
$$2x_1 - 17x_2 + 2x_3 + x_4 = 24$$
$$3x_1 + x_2 - 20x_3 + 7x_4 = 63$$
$$x_1 + 4x_2 + 3x_3 - 23x_4 = -73.$$

Rearranging these equations to apply the Gauss–Siedel iteration yields

$$x_1^{(k+1)} = (-34 - x_2^{(k)} - 5x_3^{(k)} - 3x_4^{(k)})/(-19)$$
$$x_2^{(k+1)} = (24 - 2x_1^{(k+1)} - 2x_3^{(k)} - x_4^{(k)})/(-17)$$
$$x_3^{(k+1)} = (63 - 3x_1^{(k+1)} - x_2^{(k+1)} - 7x_4^{(k)})/(-20)$$
$$x_4^{(k+1)} = (-73 - x_1^{(k+1)} - 4x_2^{(k+1)} - 3x_3^{(k+1)})/(-23)$$

and taking the initial guess

$$\mathbf{x}^{(0)} = (0 \ \ 0 \ \ 0 \ \ 0)^{\mathrm{T}}$$

leads to the estimates

$$x_1^{(1)} = 34/19 = 1.78947$$
$$x_2^{(1)} = (24 - 2(1.78947))/(-17) = -1.20124$$
$$x_3^{(1)} = (63 - 3(1.78947) + 1.20124)/(-20) = -2.94164$$
$$x_4^{(1)} = (-73 - 1.78947 - 4(-1.20124) - 3(-2.94164)/(-23) = 2.65911.$$

Continuing the iteration to obtain four-decimal-place accuracy leads to the successive estimates

x_1	x_2	x_3	x_4
1.37200	-1.44001	-2.08551	2.71110
1.59294	-1.31024	-2.02769	2.75082
1.62125	-1.29777	-2.00891	2.75667
1.62777	-1.29445	-2.00572	2.75795
1.62899	-1.29385	-2.00506	2.75819
1.62923	-1.29373	-2.00494	2.75824
1.62928	-1.29371	-2.00491	2.75825
1.62929	-1.29371	-2.00491	2.75825

i.e. to four decimal places the answer is

$$x_1 = 1.6293, \quad x_2 = -1.2937, \quad x_3 = -2.0049, \quad x_4 = 2.7583.$$

6.2.3 Successive over-relaxation (S.O.R.)

Algorithm 6.3. Simple form of Gauss–Siedel

The matrix A is augmented by making b its $(n+1)$th column, and is best arranged so that diagonal elements dominate. The tolerance is Tol.

```
read(n, Tol);
FOR i:= 1 TO n DO
   BEGIN
   FOR j:= 1 TO n DO read(A[i,j]);
   read(b[i])
   END;
FOR i:= 1 TO n DO x[i]:= 0;
Loop:= true;
WHILE Loop DO
   BEGIN
   Loop:= false;
   FOR i:= 1 TO n DO
      BEGIN
      CorrTerm:= b[i];
      FOR j:= 1 TO n DO CorrTerm:= CorrTerm - A[i,j]*x[j];
      CorrTerm:= CorrTerm/A[i,i];
      x[i]:= x[i] + CorrTerm;
      IF abs(CorrTerm) > Tol THEN Loop:= true
      END
   END;
write(x).
```

Successive over-relaxation is a modification of the Gauss–Siedel method. More specifically, we will see that the Gauss–Siedel method is a special case of S.O.R. In the Gauss–Siedel method, $x_i^{(k+1)}$ was calculated according to equation (6.39). Adding and subtracting $x_i^{(k)}$ to the right-hand side of equation (6.39) yields

$$x_i^{(k+1)} = x_i^{(k)} + \left\{ b_i - \sum_{l=1}^{i-1} a_{il}x_l^{(k+1)} - \sum_{l=i}^{n} a_{il}x_l^{(k)} \right\} \Big/ a_{1i}. \qquad (6.41)$$

Notice that the added $x_i^{(k)}$ appears to the right of the equals sign and the subtracted $x_i^{(k)}$ is included by changing the lower limit in the second summation. With the Gauss–Siedel scheme written in this way we see that the quantity

$$\left\{ b_i - \sum_{l=1}^{i-1} a_{il}x_l^{(k+1)} - \sum_{l=i}^{n} a_{il}x_l^{(k)} \right\} \qquad (6.42)$$

is the amount by which x_i changes between two successive stages of the iteration. Thus (6.42) is the error or correction term that reduces to zero when the scheme converges. IN S.O.R. this correction term is enhanced by scaling it by a factor ω, called the relaxation factor, yielding

$$x_i^{(k+1)} = x_i^{(k)} + \omega \left\{ b_i - \sum_{l=1}^{i-1} a_{il}x_l^{(k+1)} - \sum_{l=i}^{n} a_{il}x_l^{(k)} \right\} \Big/ a_{ii}, \qquad (6.43)$$

where $\omega > 1$ for S.O.R. The determination of an optimum value for ω to give the most rapid convergence requires an extensive knowledge of linear algebra and is beyond the scope of this book. The value of ω normally lies between 1 and 1.5. The application of

S.O.R. is no more difficult than the application of the Gauss–Siedel scheme which was demonstrated in Example 6.3.

6.2.4 Convergence

For a general system of linear equations the derivation of criteria for the convergence of an iterative scheme is beyond the scope of this book. However we mention two conditions that, when satisfied, will guarantee convergence of both the Jacobi and the Gauss–Siedel schemes. These schemes will converge if the $n \times n$ matrix of coefficients $\mathbf{A}$ has the following two properties:

(i) The matrix $\mathbf{A}$ does not contain a $p \times q$ sub-matrix of zeros with $p + q = n$. Such a matrix is termed irreducible.
(ii) The magnitude $\mathbf{A}$ of each diagonal element of $\mathbf{A}$ is at least as large as the sum of the magnitudes of the other elements in its row, and, in at least one row, is larger than the sum. Such a matrix is said to possess diagonal row dominance.

6.2.5 The iterative improvement procedure for removing round-off error

Unless a system of linear equations can be solved using integer arithmetic, the computed solution of the system will almost always be approximate because of round-off error. For some linear systems, round-off error can present a major problem. Here we describe a method for detecting round-off error and an iterative technique for reducing the magnitude of round-off error.

Let $\bar{\mathbf{x}}$ denote a computed solution of the system of linear equations (6.1). If we substitute $\bar{\mathbf{x}}$ back into the system of equations (6.1) we may find that, because of round-off error, the equation is not satisfied, i.e. $\mathbf{A}\bar{\mathbf{x}}$ is not equal to $\mathbf{b}$. Thus, defining $\bar{\mathbf{b}}$ by $\bar{\mathbf{b}} = \mathbf{A}\bar{\mathbf{x}}$, any difference between $\mathbf{b}$ and $\bar{\mathbf{b}}$ indicates the presence of round-off error.

The method that we present shows how to generate a sequence of approximate solution vectors $\mathbf{x}^{(0)}, \mathbf{x}^{(1)}, \mathbf{x}^{(2)}, \ldots$ to (6.1) with the aim that

$$\mathbf{x}^{(k)} \to \mathbf{x} \quad \text{as} \quad k \to \infty. \tag{6.44}$$

In practice, members of the sequence are computed until

$$\max_{1 \leqslant i \leqslant n} |x_i^{(k+1)} - x_i^{(k)}| < \varepsilon, \tag{6.45}$$

where ε is some prescribed tolerance that will determine the number of iterations, $k + 1$.

From any computed solution vector $\mathbf{x}^{(k)}$ the method supplies a new computed solution vector $\mathbf{x}^{(k+1)}$. To derive the method we substitute $\mathbf{x}^{(k)}$ back into the system of equations (6.1) and calculate $\mathbf{A}\mathbf{x}^{(k)}$. Because of round-off error the column vector

$$\mathbf{b}^{(k)} = \mathbf{A}\mathbf{x}^{(k)}, \tag{6.46}$$

which defines $\mathbf{b}^{(k)}$, can differ from $\mathbf{b}$, the right-hand side of (6.1). Forming the difference between (6.1) and (6.44) gives

$$\mathbf{A}(\mathbf{x} - \mathbf{x}^{(k)}) = \mathbf{b} - \mathbf{b}^{(k)}$$

or

$$\mathbf{A}\delta\mathbf{x}^{(k)} = \delta\mathbf{b}^{(k)}, \tag{6.47}$$

where

$$\delta x^{(k)} = x - x^{(k)} \qquad\qquad (6.48)$$

is called the correction vector and

$$\delta b^{(k)} = b - b^{(k)} \qquad\qquad (6.49)$$

is called the residual vector. In particular we find that the correction vector satisfies a system of equations with the same matrix of coefficients as the original system of equations (6.1).

The correction vector $\delta x^{(k)}$ can be calculated by solving the system of equations (6.47) and by rearranging (6.48),

$$x = x^{(k)} + \delta x^{(k)}, \qquad\qquad (6.50)$$

we see that $x + \delta x$ gives a new solution vector. However, because $\delta x^{(k)}$ will also contain round-off error, (6.50) will not be satisfied exactly, i.e. in general $x^{(k)} + \delta x^{(k)}$ will differ from the exact solution vector x and should be regarded as a new approximation to x. Thus, instead of (6.50), we write

$$x^{(k+1)} = x^{(k)} + \delta x^{(k)}. \qquad\qquad (6.51)$$

In summary, the iterative improvement procedure leads to the estimate $x^{(k+1)}$ of the solution vector x via

(i) calculating $b^{(k)} = Ax^{(k)}$,
(ii) calculating the residual vector $\delta b^{(k)} = b - b^{(k)}$,
(iii) solving $A\delta x^{(k)} = \delta b^{(k)}$ to determine the correction vector $\delta x^{(k)}$
(iv) calculating the new estimate $x^{(k+1)} = x^{(k)} + \delta x^{(k)}$.

The initial estimate, $x^{(0)}$, to the solution vector is obtained by solving the original system of equations $Ax = b$ by triangular factorization for example. Triangular factorization is a particularly good method to use since the bulk of the calculation effort is involved in forming the matrices L and U (see section 6.1.3). Compared to other methods this reduces the computational effort required to calculate the residual vectors $\delta x^{(0)}, \delta x^{(1)}, \ldots$. Details of the application of the iterative scheme are summarized in the following algorithm.

Algorithm 6.4. The iterative improvement procedure

LUsolve is a procedure to solve a matrix equation based on the LU method and MatProd a procedure to matrix multiply.

```
read(n, Tol);
FOR i:= 1 TO n DO
   BEGIN
   FOR j:= 1 TO n DO read(A[i, j]);
   read(b[i])
   END;
```

LUsolve(A, x, b, n);
Loop:= true;
WHILE Loop DO
  BEGIN
    MatProd(A, x, bk, n);
    FOR $i := 1$ TO n DO Del$b[i] := b[i] - bk[i]$;
    LUsolve(A, Del$x$, Delb, n);
    FOR $i := 1$ TO n DO $xk[i] := x[i] + Delx[i]$;
    FOR $i := 1$ TO n DO
      IF abs($xk[i] - x[i]$) > Tol THEN Loop:= true;
    IF Loop THEN
      FOR $i := 1$ TO n DO $x[i] := xk[i]$
  END
write(x).

Algorithm 6.2 (triangular factorization) could be used for steps (2) and (6). Thus when Algorithm 6.4 is coded to form a computer program, Algorithm 6.1 should be coded into a subroutine for this program.

In some problems, when round-off is a major source of error, it may be necessary to apply iterative improvement and work to a large number of significant figures, i.e. use double-precision arithmetic.

EXERCISES 6

1. Use Gaussian elimination to solve the following systems of equations:

(a)
$$2x_1 + 2x_2 + 3x_3 = 4$$
$$4x_1 - 2x_2 + x_3 = 9$$
$$x_1 + 5x_2 + 4x_3 = 3$$

(b)
$$x_1 + 4x_2 + x_3 = 5$$
$$6x_1 + 2x_2 + 3x_3 = -1$$
$$2x_1 - x_2 + 4x_3 = 7$$

(c)
$$9x_1 + 3x_2 + 4x_3 = 11$$
$$2x_1 + 7x_2 + 2x_3 = 8$$
$$3x_1 + x_2 + 6x_3 = 12$$

2. Use Gaussian elimination with partial pivoting to solve the following systems of equations:

(a)
$$x_1 - 2x_2 + 3x_3 + 9x_4 = 5$$
$$3x_1 + 10x_2 + 4x_3 + 2x_4 = 7$$
$$11x_1 + 5x_2 + 9x_3 + 2x_4 = 13$$
$$2x_1 + 3x_2 + 7x_3 + 6x_4 = 11$$

(b)
$$x_1 + 2x_2 + 11x_3 + 3x_4 = 10$$
$$10x_1 - x_2 - 2x_3 + 3x_4 = 7$$
$$3x_1 - 3x_2 + 4x_3 + 9x_4 = -6$$
$$9x_2 + 3x_3 - x_4 = 14.$$

3. Use Gaussian elimination with scaled partial pivoting to solve the system of equations

$$57x_1 + 45.6x_2 + 5.7x_3 + 11.4x_4 = 119.7$$
$$0.8x_1 - 0.2x_2 + 3x_3 + 2.4x_4 = 0.8$$
$$20x_1 - 12.5x_2 - 60x_3 - 7.5x_4 = -32.5$$
$$x_1 + 2x_2 - 15x_3 + 2x_4 = -12 .$$

4. Find lower and upper triangular matrices L, U, with unit elements on the diagonal of L, such that $LU = A$, where A is the matrix of coefficients in Exercises 1(a), 1(b) and 1(c). Hence solve the systems of equations in these questions using forward and backward substitution.

5. Solve the system of equations $Ax = b$, where A is the matrix of coefficients in Exercise 1(c), for each of

   (a) $b = (10 \quad 7 \quad 13)^T$
   (b) $b = (11 \quad 8 \quad 10)^T$
   (c) $b = (16 \quad 2 \quad 7)^T$

6. If A is the diagonally dominant matrix obtained by re-ordering the rows of the matrix of coefficients in Exercise 2(a), solve the system of equations $Ax = b$ for each of

   (a) $b = (-2 \quad 5 \quad 8 \quad 5)^T$
   (b) $b = (11 \quad 9 \quad 6 \quad 7)$
   (c) $b = (4 \quad -5 \quad -2 \quad 13)^T$.

7. If X is the inverse of the $n \times n$ matrix A then $AX = I$, where I is the identity matrix. Thus, denoting column i of X by x_i and column i of I by b_i, column i of the inverse matrix can be found by solving

$$Ax_i = b_i$$

for $i = 1(1)n$. Use an efficient method to find the inverses of the matrices

   (a) 9 3 4   (b) 10 −1 −2 −3
       2 7 2   0   9   3 −1
       3 1 6   1   2 11   3
                           3 −3   4   9

8. Solve the following tridiagonal systems of equations:

   (a)
   $$3x_1 + x_2 = 4$$
   $$2x_1 + 5x_2 + x_3 = 3$$
   $$x_2 + 4x_3 + 2x_4 = 5$$
   $$x_3 + 3x_4 = 6$$

(b)
$$6x_1 + 2x_2 \qquad\quad = -2$$
$$3x_1 + 7x_2 - x_3 = 3$$
$$2x_2 + 5x_3 + x_4 = 4$$
$$x_3 + 4x_4 = 5$$

(c)
$$-7x_1 + 3x_2 \qquad\quad = -6$$
$$-x_1 + 8x_2 - 3x_3 = 5$$
$$x_2 + 6x_3 + 2x_4 = 4$$
$$2x_3 - 5x_4 - 2x_5 = 8$$
$$4x_4 + 9x_5 - 2x_6 = 7$$
$$3x_5 + 7x_6 = 4.$$

9. Solve the following systems of equations using Jacobi's method:

(a)
$$2843x_1 + 1.326x_2 + 9.841x_3 = 5.643$$
$$8.673x_1 + 1.295x_2 - 3.215x_3 = 3.124$$
$$0.173x_1 - 7.724x_2 + 2.832x_3 = 1.694$$

(b)
$$10x_1 + 2x_2 + 3x_3 = 17$$
$$x_1 + 8x_2 - x_4 = -3$$
$$2x_1 + 7x_3 + 11x_4 = 30$$
$$2x_1 + 10x_3 + 2x_4 = 27.$$

10. Solve the system of equations in Exercise 9 using

(a) the Gauss–Siedel method,
(b) S.O.R. with a relaxation factor of 1.2.

Experiment with different relaxation factors between 1 and 1.5 to find the most rapid convergence.

11. Solve the system of equations

$$6.8259 \times 10^{-8}\, x_1 + 1.6071 \times 10^{-3} x_2 \qquad\qquad\qquad = 2.5597 \times 10^{-8}$$
$$1.0970 \times 10^{-10} x_1 + 9.0402 \times 10^{-4} x_2 + 3.4130 \times 10^{-8} x_3 = -2.8739 \times 10^{-8}$$
$$9.0402 \times 10^{-4}\, x_1 \qquad\qquad\qquad + 1.6071 \times 10^{-3} x_3 = -8.8784 \times 10^{-4}$$

using S.O.R. with a relaxation factor equal to 1.2. Also use a relaxation factor of 0.8 (when the relaxation factor is less than unity the process is termed under-relaxation). Experiment with the order of the equations to obtain the most rapid convergence.

12. The first computed solution of the system of equations

$$2.71 \quad 1.63 \quad 0.32 \quad x_1 = 4.6$$
$$4.11 \quad 2.44 \quad 0.19 \quad x_2 = 6.19$$
$$2.69 \quad 1.64 \quad 0.36 \quad x_3 = 4.705$$

is $x_1 = 0.4912$, $x_2 = 1.5150$, $x_3 = 2.4968$. Use the iterative improvement procedure to check or improve this solution and obtain four-decimal-place accuracy.

Appendix

HOMOGENEOUS LINEAR DIFFERENCE EQUATIONS

This appendix does not give a complete description of the solution of difference equations. It is only intended to aid the understanding of those difference equations that arise in section 5.1.5.

Difference equations are equations that involve or can be expressed in terms of finite differences such as Δw_n, $\Delta^2 w_n$, etc. (see section 3.1). Whereas the function $y(x)$ is usually defined over some continuous range of values of x, the function w_n is only defined for integer values of n. Thus the graph of w_n is a discrete set of points and w_n is termed a discrete function. Recall that, in Chapter 5, w_n is a numerical approximation for $y(x_n)$ and that $x_{n+1} = x_n + h$.

Via equations (3.23), (3.24), (3.25), finite differences can be expressed in terms of function values, e.g. $\Delta^2 w_n = w_{n+2} - 2w_{n+1} + w_n$. Consequently difference equations can be expressed in terms of function values, and it is in this form that they usually appear. A simple example of a difference equation is

$$w_{n+2} = 5w_{n+1} - 6w_n + (n^3 - 6n + 7). \tag{A.1}$$

Difference equations are said to be linear when there are no cross-products or powers of $w_n, w_{n+1}, \ldots$ in the equation. Thus the adjective relates to the dependent variable w_n and not to the independent variable n. Equation (A.1) is a linear difference equation, even though it is non-linear in n. The most general form of a linear difference equation, relating $p + 1$ successive members of the sequence $\{w_k\}$, can be expressed in the form

$$a_0 w_{n+p} + a_1 w_{n+p-1} + \cdots + a_p w_n = \phi(n). \tag{A.2a}$$

where $\phi(n)$ is an arbitrary function of n. In general, the coefficients a_i, $i = 0(1)p$. are functions of n but we will only consider the case when they are constants. Using the shift operator E (see section 3.1) equation (A.2a) can be written

$$(a_0 E^p + a_1 E^{p-1} + \cdots + a_p)w_n = \phi(n). \tag{A.2b}$$

Equation (A.2) has order p, the order of a difference equation being defined as the difference between the highest and lowest suffices in the equation.

A difference equation is said to be homogeneous when every term in the equation contains the dependent variable w. Otherwise the equation is said to be non-homogeneous. Equation (A.1) is non-homogeneous. Equation (A.2) is homogeneous or non-homogeneous depending on whether or not $\phi(n)$ is identically zero. For section 5.1.5 we need only consider homogeneous linear difference equations. Thus we restrict attention to the equation

$$a_0 w_{n+p} + a_1 w_{n+p-1} + \cdots + a_p w_n = 0 \tag{A.3a}$$

or

$$(a_0 E^p + a_1 E^{p-1} + \cdots + a_p) w_n = 0, \tag{A.3b}$$

where the coefficients a_i, $i = 0(1)p$. are constants.

The general solution of a second-order linear difference equation contains two arbitrary constants. This is similar to differential equation theory. To justify this, consider equation (A.1). If α and β are the numerical values of w_0 and w_1 for this equation, numerical values of w_2, w_3, ... can be calculated in turn by setting $n = 0, 1, \ldots$ in (A.1). The complete solution of this second-order linear difference equation depends on the two values of α and β. These values are usually supplied by boundary or initial conditions. Similarly the general solutions of equations (A2), (A3), which have order p, contain p arbitrary constants.

To obtain the general solution of the homogeneous linear difference equation (A.3) we initially consider the first-order equation

$$(a_0 E + a_1) w_n = 0, \tag{A.4}$$

which can be obtained by setting $p = 1$ in (A.3b). This equation can be written

$$w_{n+1} = \left(-\frac{a_1}{a_0} \right) w_n = \lambda w_n, \tag{A.5}$$

where $\lambda = -a_1/a_0$. Observe that λ is a root of the equation

$$a_0 \lambda + a_1 = 0 \tag{A.6}$$

and compare the form of this equation with (A.4). If $w_0 = A$, where A is an arbitrary constant, setting $n = 0, 1, 2, \ldots$ in (A.5) gives

$$w_1 = \lambda A, \quad w_2 = \lambda^2 A, \quad w_3 = \lambda^3 A, \ldots.$$

Thus in general we have

$$w_n = A \lambda^n, \tag{A.7}$$

i.e. this is the general solution of the first-order equation (A.4), where A is an arbitrary constant and λ is the root of equation (A.6).

We now consider the second-order equation

$$a_0 w_{n+2} + a_1 w_{n+1} + a_2 w_n = 0 \tag{A.8a}$$

or

$$(a_0 E^2 + a_1 E + a_2) w_n = 0, \tag{A.8b}$$

which can be obtained by setting $p = 2$ in equations (A.3a), (A.3b). This equation can be written

$$\left(E^2 + \frac{a_1}{a_0}E + \frac{a_2}{a_0} \right) w_n = 0$$

or, factorizing,

$$(E - \lambda_1)(E - \lambda_2)w_n = 0, \tag{A.8c}$$

where λ_1, λ_2 are the roots of the quadratic equation

$$a_0\lambda^2 + a_1\lambda + a_2 = 0. \tag{A.9}$$

Equation (A.8) will be satisfied if either of the first-order equations

$$(E - \lambda_1)w_n = 0, \quad (E - \lambda_2)w_n = 0$$

is satisfied. Thus, using equation (A.7), each of $A_1\lambda_1^n$ and $A_2\lambda_2^n$ is a solution of equation (A.8), and it is easily verified by substitution that

$$w_n = A_1\lambda_1^n + A_2\lambda_2^n \tag{A.10}$$

also satisfies equation (A.8). Further, this latter solution contains two arbitrary constants, A_1 and A_2, and is the general solution of (A.8), provided λ_1 and λ_2 are distinct.

To summarize, we have shown that (A.10) is the general solution of (A.8), provided λ_1 and λ_2 are distinct roots of equation (A.9). The case $\lambda_1 = \lambda_2$ has to be treated separately but is not needed here.

In a similar manner it can be shown that the pth-order homogeneous equation (A.3) has general solution

$$w_n = \sum_{i=1}^{p} A_i\lambda_i^n \tag{A.11}$$

provided that λ_i, $i = 1(1)p$, are distinct roots of the polynomial equation

$$a_0\lambda^p + a_1\lambda^{p-1} + \cdots + a_p = 0. \tag{A.12}$$

In particular, observe the similarity in the forms of equations (A.3b), and (A.12). Equation (A.12) is given the special name auxiliary equation because of its key role in obtaining the solution. Equations (A.6) and (A.9) are also auxiliary equations corresponding to $p = 1$ and $p = 2$ respectively in (A.12).

Index

Adams-Bashforth formulae, 91
Adams-Moulton formulae, 93
Auxiliary equation, 146
Averaging operator μ, 37

Bessel's formula, 56
Birge-Vieta method, 24
Bisection method, 15

Central difference interpolation formulae, 52
Choleski's method, 129
Closed and open integration formulae, 82
Convergence, 7, 139
  test for, 9
  rate of, 14
  of corrector formula, 105
Correction term, 18
Correction vector, 140
Corrector formulae, 89
Cramer's rule, 121
Crout's method, 129

Difference operators, 32
  shift operator E, 36
  averaging operator μ, 37
Divided differences
  Newton's divided difference formula, 45
  Newton's divided difference polynomial, 47
Doolittle's method, 129

E (shift operator), 36
Error, 3
  error propagation-stability, 106
  estimation of truncation error, 103
  global truncation error, 66
  local truncation error, 66
  truncation error, 75
Extrapolation, 31

Finite differences and difference operators, 32

Gauss forward and backward interpretation
  formulae, 54
Gaussian elimination, 122

Gauss-Siedel method, 136
Global truncation error, 66
Gregory-Newton forward and backward
  interpretation formulae, 40

Homogeneous linear difference equations, 144

Integration
  numerical, 63
  open and closed integration formulae, 82
Ill-conditioned problems, 107
Interpolation
  linear, 30
  formulae, 40
    central difference, 52
    Gauss forward and backward, 54
    Gregory-Newton, 40
Iteration by algebraic transformation, 7

Jacobi's method, 135

Linear interpolation, 30
Local truncation error, 66

Mean-value theorem, 3
Milne's formula, 83
Milne's open formulae, 102
Milne-Simpson method, 102
Modelling, numerical/computational, 1
Multi-step methods, 89

Newton-Cotes formulae, 64
Newton's divided difference formula, 45
Newton's divided difference polynomial, 47
Newton-Raphson method, 17
Numerical/computational modelling, 1
Numerical integration, 63

Open and closed integration formulae, 82
Ordinary differential equations, numerical
  solution of, 87

Parasitic solutions, 108
Picard's method, 96
Pivot, pivotal equation, pivoting, 123
Polynomial equations, real roots of, 22
Predictor-corrector methods, 88

Predictor formulae, 88

Real roots of polynomial equations, 22
Recurrence relation, 8
Relaxation factor, 138
Remainder theorem, 22
Residual vector, 140
Rollé's theorem, 3
Romberg integration, 78
Runge-Kutta formulae and methods, 113

Sequence generating function, 9
Shift operator E, 36
Simpson's closed formula, 102
Simpson's rule, 70
Stability-error propagation, 106

Stability polynomial, 110
Stirling's formula, 55
Successive over-relation (S.O.R.), 137
Synthetic division, 23

Taylor, 14
  expansion, 5
  series, 5
  series method, 94
Three-eighths rule, 77
Trapezium rule, 66
Triangular factorization, 128
Triangular systems, 121
Truncation error, 66, 75
  local, 66
  global, 66

Mathematics and its Applications

Series Editor: G. M. BELL, Professor of Mathematics, King's College London (KQC), University of London

Faux, I.D. & Pratt, M.J.	Computational Geometry for Design and Manufacture
Firby, P.A. & Gardiner, C.F.	Surface Topology
Gardiner, C.F.	Modern Algebra
Gardiner, C.F.	Algebraic Structures: with Applications
Gasson, P.C.	Geometry of Spatial Forms
Goodbody, A.M.	Cartesian Tensors
Goult, R.J.	Applied Linear Algebra
Graham, A.	Kronecker Products and Matrix Calculus: with Applications
Graham, A.	Matrix Theory and Applications for Engineers and Mathematicians
Graham, A.	Nonnegative Matrices and Applicable Topics in Linear Algebra
Griffel, D.H.	Applied Functional Analysis
Griffel, D.H.	Linear Algebra
Guest, P. B.	The Laplace Transform and Applications
Hanyga, A.	Mathematical Theory of Non-linear Elasticity
Harris, D.J.	Mathematics for Business, Management and Economics
Hart, D. & Croft, A.	Modelling with Projectiles
Hoskins, R.F.	Generalised Functions
Hoskins, R.F.	Standard and Non-standard Analysis
Hunter, S.C.	Mechanics of Continuous Media, 2nd (Revised) Edition
Huntley, I. & Johnson, R.M.	Linear and Nonlinear Differential Equations
Jaswon, M.A. & Rose, M.A.	Crystal Symmetry: The Theory of Colour Crystallography
Johnson, R.M.	Theory and Applications of Linear Differential and Difference Equations
Johnson, R.M.	Calculus: Theory and Applications in Technology and the Physical and Life Sciences
Jones, R.H. & Steele, N.C.	Mathematics of Communication
Jordan, D.	Geometric Topology
Kelly, J.C.	Abstract Algebra
Kim, K.H. & Roush, F.W.	Applied Abstract Algebra
Kim, K.H. & Roush, F.W.	Team Theory
Kosinski, W.	Field Singularities and Wave Analysis in Continuum Mechanics
Krishnamurthy, V.	Combinatorics: Theory and Applications
Lindfield, G. & Penny, J.E.T.	Microcomputers in Numerical Analysis
Livesley, K.	Engineering Mathematics
Lord, E.A. & Wilson, C.B.	The Mathematical Description of Shape and Form
Malik, M., Riznichenko, G.Y. & Rubin, A.B.	Biological Electron Transport Processes and their Computer Simulation
Massey, B.S.	Measures in Science and Engineering
Meek, B.L. & Fairthorne, S.	Using Computers
Menell, A. & Bazin, M.	Mathematics for the Biosciences
Mikolas, M.	Real Functions and Orthogonal Series
Moore, R.	Computational Functional Analysis
Murphy, J.A., Ridout, D. & McShane, B.	Numerical Analysis, Algorithms and Computation
Nonweiler, T.R.F.	Computational Mathematics: An Introduction to Numerical Approximation
Ogden, R.W.	Non-linear Elastic Deformations
Oldknow, A.	Microcomputers in Geometry
Oldknow, A. & Smith, D.	Learning Mathematics with Micros
O'Neill, M.E. & Chorlton, F.	Ideal and Incompressible Fluid Dynamics
O'Neill, M.E. & Chorlton, F.	Viscous and Compressible Fluid Dynamics
Page, S. G.	Mathematics: A Second Start
Prior, D. & Moscardini, A.O.	Model Formulation Analysis
Rankin, R.A.	Modular Forms
Scorer, R.S.	Environmental Aerodynamics
Smith, D.K.	Network Optimisation Practice: A Computational Guide
Shivamoggi, B.K.	Stability of Parallel Gas Flows
Stirling, D.S.G.	Mathematical Analysis
Sweet, M.V.	Algebra, Geometry and Trigonometry in Science, Engineering and Mathematics
Temperley, H.N.V.	Graph Theory and Applications
Thom, R.	Mathematical Models of Morphogenesis
Thurston, E.	Primary Mathematics: Teaching and Learning
Townend, M. S.	Mathematics in Sport
Townend, M.S. & Pountney, D.C.	Computer-aided Engineering Mathematics
Twizell, E.H.	Computational Methods for Partial Differential Equations
Twizell, E.H.	Numerical Methods, with Applications in the Biomedical Sciences
Vince, A. and Morris, C.	Mathematics for Computer Studies
Walton, K., Marshall, J., Gorecki, H. & Korytowski, A.	Control Theory for Time Delay Systems
Warren, M.D.	Flow Modelling in Industrial Processes
Wheeler, R.F.	Rethinking Mathematical Concepts
Willmore, T.J.	Total Curvature in Riemannian Geometry
Willmore, T.J. & Hitchin, N.	Global Riemannian Geometry

Numerical Analysis, Statistics and Operational Research

Editor: B. W. CONOLLY, Professor of Mathematics (Operational Research), Queen Mary College, University of London

Beaumont, G.P.	Introductory Applied Probability
Beaumont, G.P.	Probability and Random Variables
Conolly, B.W.	Techniques in Operational Research: Vol. 1, Queueing Systems
Conolly, B.W.	Techniques in Operational Research: Vol. 2, Models, Search, Randomization
Conolly, B.W.	Lecture Notes in Queueing Systems
Conolly, B.W. & Pierce, J.G.	Information Mechanics: Transformation of Information in Management, Command, Control and Communication
French, S.	Sequencing and Scheduling: Mathematics of the Job Shop
French, S.	Decision Theory: An Introduction to the Mathematics of Rationality
Griffiths, P. & Hill, I.D.	Applied Statistics Algorithms
Hartley, R.	Linear and Non-linear Programming
Jolliffe, F.R.	Survey Design and Analysis
Jones, A.J.	Game Theory
Kapadia, R. & Andersson, G.	Statistics Explained: Basic Concepts and Methods
Moscardini, A.O. & Robson, E.H.	Mathematical Modelling for Information Technology
Moshier, S.	Mathematical Functions for Computers
Oliveira-Pinto, F.	Simulation Concepts in Mathematical Modelling
Ratschek, J. & Rokne, J.	New Computer Methods for Global Optimization
Schendel, U.	Introduction to Numerical Methods for Parallel Computers
Schendel, U.	Sparse Matrices
Sehmi, N.S.	Large Order Structural Eigenanalysis Techniques: Algorithms for Finite Element Systems
Späth, H.	Mathematical Software for Linear Regression
Spedicato, E. and Abaffy, J.	ABS Projection Algorithms
Stoodley, K.D.C.	Applied and Computational Statistics: A First Course
Stoodley, K.D.C., Lewis, T. & Stainton, C.L.S.	Applied Statistical Techniques
Thomas, L.C.	Games, Theory and Applications
Whitehead, J.R.	The Design and Analysis of Sequential Clinical Trials